the **physics** of PITCHING

Learn the **mechanics, science,** and **psychology** of pitching to success

LEN SOLESKY AND **JAMES T. CAIN,** WITH **RUSTY MEACHAM**
PHOTOGRAPHY BY **BRUCE CURTIS**

MVP
BOOKS

First published in 2011 by MVP Books, an imprint of MBI Publishing Company and the Quayside Publishing Group, 400 First Avenue North, Suite 300, Minneapolis, MN 55401 USA

The information in this book is true and complete to the best of our knowledge. All recommendations are made without any guarantee on the part of the author or Publisher, who also disclaims any liability incurred in connection with the use of this data or specific details.

We recognize, further, that some words, model names, and designations mentioned herein are the property of the trademark holder. We use them for identification purposes only. This is not an official publication.

MVP Books titles are also available at discounts in bulk quantity for industrial or sales-promotional use. For details write to Special Sales Manager at Quayside Publishing Group, 400 First Avenue North, Suite 300, Minneapolis, MN 55401 USA.

To find out more about our books, visit us online at www.mvpbooks.com.

Library of Congress Cataloging-in-Publication Data

Cain, James T., 1956–
The physics of pitching : learn the mechanics, science, and psychology of pitching to success / James T. Cain and Len Solesky ; photography by Bruce Curtis.
    p. cm.
Includes index.
ISBN 978-0-7603-3850-6 (flexibound)
1. Pitching (Baseball) 2. Baseball—Physiological aspects. 3. Baseball—Psychological aspects. 4. Human mechanics. 5. Force and energy. I. Solesky, Len. II. Title.
GV871.C35 2011
796.357'22—dc22

2010045070

Editor: Adam Brunner
Design manager: Kou Lor
Designer: Simon Larkin
Layout: John Sticha

Front Cover Photos by Thomas Grim

Printed in China

Contents

Introduction

It's a familiar scene, played out in one guise or another for 80 or 90 years. A wife asks her husband, a first-time dad, what he plans to do for their new son on his very first birthday, "which is coming up in a month, you know." As the toddler takes his first plodding steps away from the coffee table, the father can only smile. For the first and perhaps the only time, he is way ahead of the birthday game. Purchased months ago, his son's gift is safely stashed away in the top of the closet: the boy's first baseball glove—full-grain cowhide brown leather, full pocket! Sure, his hand will need to grow into the glove, but the thought is there.

A few years later, after plenty of pitch-and-catch in the backyard, that young boy is trying out for his first season of Little League baseball. It's time for old dad to make a pitcher out of him, just like he was (or might have been.) Dad follows baseball in the news and watches every game he can. He cringes at the injury reports of his favorite team. He remembers his sore arm, the reason he quit playing. Naturally, the father goes to the bookstore or library, looking for some guide that will tell him how to teach his boy to pitch correctly. Years have passed since his sandlot days, so he needs some reminders and some health lessons.

Junior may never be a superstar. Junior may never attract scouts when he is still in high school, or even college. He may never be drafted. He may never play professional baseball. But that doesn't mean that Junior can't have fun or reach his potential in youth league.

Good coaching and instruction are key to instilling correct habits and techniques in ballplayers at all levels. More than two million kids play organized baseball in the United States alone, with more than 10 million participants worldwide.

Kids develop at different rates and in different stages. A gangly kid one year can easily grow into a more efficient body by the next spring. Unfortunately, a lot of amateur and high school coaches tend to pigeonhole kids who are not head-and-shoulders above the rest—and some simply are always going to be better athletes. But there's a lot to be gained by learning how to play the game—with practice.

Practice done right produces muscle memory, which is what pitching is all about. The same motion executed over and over again correctly produces strikes and outs, just as a tennis player's serve produces aces and points or a golfer's swing produces well-struck shots. If repeated incorrectly, that same motion may produce strikes for a while, but it will inevitably lead to a sore arm.

A sore arm can mean anything from mildly sprained ligaments to frayed or torn ligaments. It can also mean an interruption of the growth of the bones in the arm. The bones in the arm do not all fully connect until a child is physically mature, and that happens at a different point for every child. For most kids, the growth plate at the upper arm doesn't mature until 19. A repetitive, unnatural motion can cause the muscles involved to put unnatural stress on the growth plates, where the ossification (the forming of the bone in the growing body) can be disturbed. Obviously, that sort of thing leads to injury. And for pitchers, not surprisingly, their greatest risk for injury is in their arm.

But before we get too grim, let's remember that pitchers start every play in baseball—and that's fun. Every year, medicine takes an enormous stride, not unlike Cy Young Award winner Tim Lincecum. Doctors and scientists have studied baseball and the biomechanics of pitching, and we now know plenty about how the body works in the controlled contortions of the

Proper positioning at each stage of the delivery is essential to being an effective and healthy pitcher. This young hurler is in the balance, or posting, position as he begins his delivery to the plate.

What this book offers:

1　A simple, step-by-step path for the beginning pitcher, using language and concepts the average father and son can understand.

2　An easily understood and amply illustrated path, with photographs and explanations showing each position the athlete must take to deliver the pitch.

3　A safe course, citing modern medical practice and recent scientific studies that debunk the potentially damaging concepts of "old school" coaching.

4　An evidence-based system—a safe, efficient, and revolutionary approach to coaching a pitcher by taking advantage of modern medical advances in biomechanics, physiology, and biokinetics.

pitching motion. You can pitch well, and you can do it without hurting your arm.

It has a cushioned cork center—the pill. Around that is wrapped yarn of wool, polyester, and cotton. The cover comes from tanned white cowhide stitched together in a figure eight of red yarn. It's a "hard ball."

Baseball took its first great leap in popularity as a pastime for Union soldiers in the Civil War when they became bored with the routine of camp life. Soldiers endured long periods of inaction between fighting and marching. During these times, they embraced a game evolved from the English game of rounders, its rules more or less established by Alexander Cartwright in 1845. Ball-and-stick games are often traced back to playful shepherds of centuries past, and it's probably safe to assume that children have been making up games with sticks and balls and rocks and bits of hardened clay since the beginning of time.

Alexander Cartwright, considered the "father of baseball." *National Baseball Hall of Fame and Library*

Albert Goodwill Spalding was an original promoter and powerhouse in baseball's early days. He was a professional baseball player, manager, and co-founder of A. G. Spalding & Bros. sporting goods company. *Prints and Photographs Division, Library of Congress*

The ball used in the early nineteenth century was softer in composition and larger in size than what became regulation by the turn of the century. The ball in use by semiprofessional teams in 1860 was a dark red or brown leather (a color somewhat similar to the red ball used in the white-flannel game of cricket). This baseball was anywhere from 9 3/4 inches to 10 inches in circumference and 5 3/4 ounces in weight, as opposed to today's smaller 9 to 9 1/4 inches and 5 ounces. Sail maker Harvey Ross, of Brooklyn's Atlantic Club, made a "dead ball," with only 1 1/2 ounces of India rubber. Cobbler John Van Horn, of New York's Baltic Club, used 2 to 2 1/2 ounces of rubber. His was the "lively ball."

The home team supplied the baseballs. Talk about your home-field advantage: If you had a rock-em-sock-em lineup, you preferred the sprightlier ball. If your team played with more with finesse, then the slower ball played to your strengths.

An 1889 advertisement for the National League's official ball, manufactured by A. G. Spalding & Bros. *General Collections, Library of Congress*

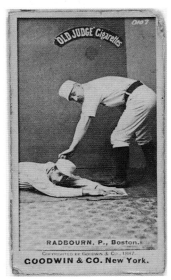

Top: Charlie Sprague wasn't much of a star, but his demonstration of the underhand delivery is featured on this early baseball card. *Prints and Photographs Division, Library of Congress*

Above: Charlie "Old Hoss" Radbourn pitched a remarkable 678 2/3 innings in the 1884 season. *Prints and Photographs Division, Library of Congress*

By 1870, truly professional teams began to take shape. If arguments over railroad gauges could be settled, then certainly some rules for baseball, a truly American game, could be reached. The keeping of statistics—batting averages, home runs, runs batted in, earned run averages, strikeouts, wins—had already begun. There had to be some way to measure ball players, to determine their relative worth on an even scale. Statistical records provided this, as well as fodder for sporting wags at the newspapers, not to mention the fans who gathered at saloons after ballgames. So for measurement's sake at least, baseballs became the standard size that they are today (although, like many people, the balls are now more tightly wound.)

The game of baseball has changed considerably throughout the years. In the 1870s, in the new National League, you could call your pitch—as the hitter, not the catcher. You could ask for a high ball or a low ball. But the pitcher didn't have to toe any rubber; he only had to throw from within a designated box, somewhat akin to the rule for a cricket hurler. Several business decisions changed the game of baseball in the late 19th century, particularly in the American Association with free agency and Sunday baseball games. But the biggest change came with the 1885 season, when the National League began to allow pitchers to throw the ball overhand. The American Association soon followed. Gloves were added a bit later, as well as a little more protective gear for catchers, but the change to overhand pitching made baseball the game we recognize today.

The baseball could be delivered faster, at more greatly varied speeds, and with unexpected curves, dips, and trifles. The spitball came into its own, until it was ruled an illegal pitch in 1920 (except for 17 designated known and proud-of-it spitballers, who were allowed grandfather exemptions for the duration of their careers). Demonstrations were set up to prove to a skeptical public that the curveball was not an optical illusion (though in many respects, it is from a batter's perspective). The pitcher could now generate greater torque and more velocity. The contortions some guys worked into to pitch the ball were often worth the price of admission on their own. Baseball had already been a spectacle, and ballplayers naturally showed off for the crowds that gathered. Now, pitchers became the star attractions.

Charlie "Old Hoss" Radbourn threw underhanded and won between 59 and 62 games in 1884 (his win statistics are debated amongst baseball historians), pitching almost every day for an incredible 678 2/3 innings. The inestimable Cy Young, who was one of the overhand-throwing newcomers, overtook Charlie down the home stretch in 1891. Chief Zimmer, who frequently caught

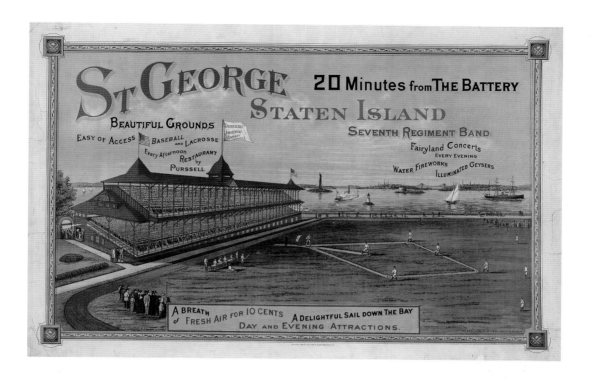

St. George Stadium on Staten Island, circa 1886. Note the relatively short distance (50 feet) between the pitcher and home plate, as well as the absence of a pitcher's mound. The distance was increased to 55 feet about a year later, with the final change to 60 feet, 6 inches and the addition of the mound coming in 1893. *Prints and Photographs Division, Library of Congress*

for "the Cyclone," is said to have put a beefsteak in his mitt for extra padding. Denton True "Cy" Young racked up 511 wins and allowed an average of 2.63 earned runs a game. Today, the best pitcher in each league is honored with an award named after him. He must have been doing something right. He probably had good mechanics, though he might have been at a loss to explain them.

Overhand pitchers like Cy Young, Amos Rusie, and Jouett Meekin threw so much harder than the underhanded gents of the 1870s that the major leagues were persuaded to move the pitching distance back 5 feet to where it is today, at 60 feet and 6 inches, according to sports historian Rob Neyer.

Rusie, "The Hoosier Thunderbolt," threw hard, fast, and wild. After being beaned by one of his pitches, Hall of Fame shortstop Hugh Jennings was unconscious for four days. Fortunately for baseball, Jennings recovered, and his near-death experience helped prompt the decision to increase the distance between the batter and the pitcher. The next year, Rusie's strikeout totals went down, but he was still a fan favorite; in fact when he held out in a contract dispute with ownership, fans and the press revolted.

Consider that at this time, Babe Ruth wasn't born yet. There were star hitters, but winning ball clubs were built around pitching (not unlike today, perhaps—despite droves of fans and sportscasters who "dig the long ball.") In 1890, solid teams only needed two

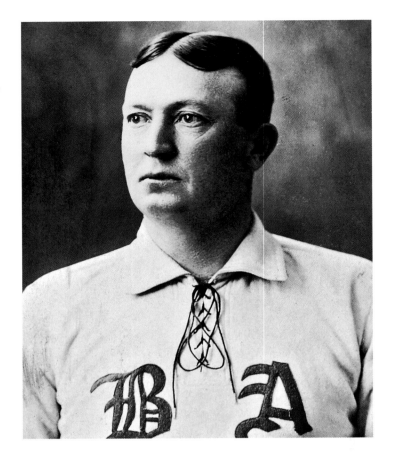

The incomparable Denton True "Cy" Young, pictured here in his Boston Americans uniform during the 1902 season. *Prints and Photographs Division, Library of Congress*

or three good pitchers. When rival leagues sprang up and raided players, pitchers were frequently the biggest stars and targets.

The early windup, not surprisingly lampooned by comics in vaudeville theaters of the day, featured double-arm pumps, high-leg kicks, and other high jinks that combined to demonstrate the pitcher's determination to put all he had into a pitch. And, if a fellow did it right, he could also distract the batter somewhat (which is all it takes) or hide the ball until just before the point of release. But in the case of the greats like Cy Young and later Walter Johnson, their bodies just naturally found and fit into the correct biomechanics. What can you say? Some guys are just able to do things better than others.

The die was cast in the early days of overhanded pitching. The method became established simply because it worked really well for the game's best pitchers—the two or three main guys a team carried in those days. Those guys were also exceptional athletes. You had to be really good to throw that many innings every year. But you also had to have pretty sound mechanics, or you'd have a rag arm in three or five years. Many did, and very few of them had careers that lasted past their 30th birthdays. And because most of them worked jobs outside of baseball, few had the luxury of off-season strength and conditioning programs.

So "the book" on pitching technique was written, set in stone to a great degree by the incredibly long careers of some managers and club owners. John McGraw ruled the New York Giants and wrote the book in the early days. The legendary Connie Mack managed the Philadelphia Athletics for 50 years. Those are just two of the famous early figures who guided the game, and there were many in between. Along with the most accomplished athletes of the era, you might think of such long-term, well-known figures as the architects of the "old school."

Cy Young managed to avoid injury in his 22 years in the major leagues. He didn't believe much in practice or excessive warming up. He did believe in control. As remarkable as his records are—throwing as much as he did, as hard as he is reported to have thrown, without getting hurt—you have to think that his delivery was nearly flawless and consistent.

"I aimed to make the batter hit the ball, and I threw as few pitches as possible. That's why I was able to work every other day," said Young. And as he aged and his fastball lost some velocity,

Chicago White Sox pitcher Jim Scott's windup was indicative of the acrobatics displayed by early overhand pitchers.
Getty Images / Chicago History Museum

he relied on a well-placed curve. On this, he noted, "Any young player who has good control will become a successful curve pitcher long before the pitcher who is endeavoring to master both curves and control at the same time. The curve is merely an accessory to control."

Greg Maddux followed the same method a century later. He learned the old-fashioned way—through skill and a few good coaches—but he also had an innate ability to transfer energy from his legs through his trunk and into his shoulder. Maddux also used his intelligence to become a successful pitcher.

A baseball pitcher's delivery is an orchestration of movement so violent and concentrated that it would outrace a Ferrari on a drag strip. Nothing else in sport accelerates so quickly from a resting point. It's all in the twist, the torque generated by the torso as it shifts the kinetic energy from the legs to the back, thrusting into the shoulder muscles. It might sound like a commercial for a sports car, but this is the power of your body.

In recent years, scientists, doctors, physical therapists, and conditioning specialists have studied how the pitching motion works. They know how force is generated, and where the body

Manager John McGraw (center) had the luxury of two stellar, Hall of Fame pitchers on his New York Giants dynasty of the early 20th century: Christy Mathewson (left) and Joe "Iron Man" McGinnity (right). Mathewson ranks as one of the game's all-time greats, with 373 career wins and 79 shutouts, while McGinnity, a two-time 30-game winner, earned his nickname by leading the league in innings pitched four times between 1900 and 1904. *National Baseball Hall of Fame and Library*

parts have to be for the machine (that is the pitching body) to work best. They have studied how to heal it when it breaks down.

One of the biggest problems is that the pitcher's delivery is at once violent, concentrated, and incredibly delicate. As with any machined tool, any variation from the mean, any slip of the axis, results in wear and tear on specific parts. If your left leg hurts, you will compensate for it with your right, and that can throw off your shoulder as well; to pitch correctly you must pitch as a whole. Every piece counts, and sometimes we don't even know when we're missing one, because we discount a minor injury or ache.

What this book promotes is a complete approach, based on how the body generates energy and pushes it into fastballs. We're going to focus on the importance of balance and where to find it. We're going to help you become a complete pitcher, no matter how young, old, or inexperienced.

Intelligence, focus, skill, and strength combined to make Greg Maddux a fearsome pitcher throughout this career.
Courtesy Ronald C. Modra

So why this book, why right now? Because too many kids are hurting their arms throwing unnaturally.

According to a study reported by Dr. James Whiteside, Dr. James Andrews, and Glenn Fleisig, PhD, in *The Physician and Sports Medicine* journal in 1999, "Repeated throwing, especially pitching, in skeletally immature athletes can produce elbow injuries that threaten the growth plate." Yet, despite youth leagues' awareness and rules about pitch counts, young pitchers continue to be overused. They get hurt, or their arms wear out prematurely.

Part of the fault rests with coaching that is often limited by an inadequate understanding of the pitching motion and pitching kinetics, or the force a pitcher generates to deliver the baseball. Part of the problem, of course, is overuse. But another big component is that players are throwing with just their arms. If a young pitcher learns to use his whole body correctly, he will throw harder and more consistently—and more importantly for pitchers of any age—without as much chance of injury. We say "chance" because every sport carries the chance of injury.

Shutterstock

Chapter 1

The Numbers

If you've picked this book up, you or your child is most likely among the millions of participants in youth baseball across the country. Almost a quarter of those kids are pitchers at some point in their playing career. Coaches will try almost anybody in some games. And doesn't every kid, deep in his heart, want to toss the old pill?

Founded by the famous orthopedic surgeon Dr. James Andrews, American Sports Medicine Institute is a nonprofit institute dedicated to injury prevention, education, and research in orthopedics and sports medicine. Based in Birmingham, Alabama, ASMI reports that "As many as 58 percent of children and adolescents between the ages of 11 and 18 experience elbow injuries during or after pitching in organized games."

Also the founder of Andrews Sports Medicine & Orthopaedic Center, Dr. James Andrews has repaired several famous athletes, including Michael Jordan, John Smoltz, and Jack Nicklaus. Another of Dr. Andrews's studies found an injury incidence of 40 percent in nearly 200 pitchers, 9 to 12 years old, who were followed for a year.

ASMI notes that the number of children who have developed serious shoulder and elbow injuries has increased sixfold in recent years.

According to several published studies, as many as 45 percent of pitchers under 12 years old complain of chronic elbow pain. At the high school level, nearly 6 in every 10 pitchers suffer chronic elbow pain.

Those are pretty high numbers, aren't they? Shouldn't we able to bring it down a little bit? Dr. Andrews's center is working pretty hard at that, especially with education. Their website, www.andrewscenters.com, has a wealth of information.

Overuse is a big problem contributing to these statistics. Youth league coaches are too tempted to use their best pitcher

Serious shoulder and elbow injuries have become increasingly common among young pitchers in recent years, largely the result of improper mechanics. This youth league pitcher shows good form and hand separation as he gets into the power slot position.

as many times as they can, even as many leagues have begun to establish rules limiting innings pitched per week. But some kids are playing on more than one team in more than one league, so it's up to their parents to keep track of their child's arm health. And kids play year-round, not just in summer months, leaving little time for rest and recuperation.

For instance, a study published by ASMI documented that injured baseball players (those requiring elbow or shoulder surgery) went to four times as many showcases as those who were in the healthy control group. Of course, showcase ball players are the better players and are more likely to play in several leagues. This isn't a bad thing (you may get more opportunities to take the mound)—if you don't overuse your arm. Using ice is always a good idea after a start, too, to keep inflammation down.

In 2008, Dr. Andrews and his colleagues conducted a study comparing 95 high school pitchers who required surgery on either an elbow or shoulder with 45 pitchers without injury. They found that those who pitched for more than eight months per year were 500 percent more likely to be injured, while those who pitched more than 80 pitches per game increased their injury risk by 400 percent. Pitchers who continued pitching despite having arm fatigue were 3,600 percent more likely to do serious damage to their arm.

A lot of the difficulty arises from the immaturity of the young ball player. His body just isn't ready. Between the ages of 8 and 14, boys are still maturing physically. In the elbow, that means the growth plates of the bone are weaker than the tendons. The bone plates have yet to ossify, or grow together. This places added stress on the tendons, stress that down the road could lead to what's called Tommy John surgery, or tendon replacement surgery.

Childhood Injuries

- Among pitchers under 12 years old, as many as 45 percent complain of chronic elbow pain, according to several published studies.

- At the high school level, nearly 6 in every 10 pitchers suffer chronic elbow pain.

- As many as 58 percent of children and adolescents between the ages of 11 and 18 experience elbow injuries during or after pitching in organized games.

- Another survey found an injury incidence of 40 percent in nearly 200 pitchers, 9 to 12 years old.

Brian Pendergast demonstrates a good square position, facing the target. His head is in a straight line over the belly button. His plant foot is flat and secure on the ground, and his shoulder has great external rotation.

Make no mistake. Pitching can lead to arm pain. Pitching a baseball is the most unnatural activity in sport. It involves quick specific movements and is the fastest moving motion in sports.

The muscles' contractile strength and flexibility contribute to kinetic energy in the pitching motion. The elastic qualities of muscles can be strengthened to increase their kinetic capabilities. In successful pitchers like Tim Lincecum and Justin Verlander, the body acts as a turbine, generating energy from the legs, through the pelvis, to the rotation of the trunk of the body to the slingshot of the arm. Energy works up the body and is unleashed into the long stride and trunk rotation.

Pitching is the most violent motion in sport. Glenn Fleisig, research director at American Sports Medicine Institute, estimates that about 80 Newton meters of torque act on a top pitcher's elbow when he throws a fastball. His experiments with cadavers show that an average person's ulnar collateral ligament snaps at 80 Newton meters. Fleisig has also found that, when the hardest thrower rotates his pelvis toward home plate, his trunk follows less than a tenth of a second later. Those numbers, of course, apply to grown men. But they signify the enormous stressors placed on a youth's arm when he is learning to pitch.

As we've said, a young child is still growing, and the muscles, bones, and tendons have not fully developed. So a youth is particularly vulnerable to elbow and shoulder injury, not just from throwing overhand too much, but from throwing incorrectly. You can limit a child's pitch count, but if his mechanics are bad, he can still hurt his arm.

All that said, the best way to avoid injury is to pitch the right way from the start, using your body and not just your strong arm. That is what we aim to teach you in this book.

Nicknamed "The Freak," Giants ace Tim Lincecum stands only 5-foot-10, but his fastball can reach 98 miles per hour. His velocity comes from sharp mechanics that enable a 7 1/2 foot stride, which is 129 percent of his height. This approach helped Lincecum win two Cy Young Awards in his first three major league seasons. *Courtesy Ronald C. Modra*

Justin Verlander by the Numbers

For Detroit Tigers right-hander Justin Verlander to throw his fastball 100 miles per hour, his body undergoes an almost-violent series of motions.

His arm's internal average rotation is 1,100 rpm.

Rotating at 1,500 rpm, the ball has almost 40 pounds of force.

As he turns to the plate his trunk must rotate at 200 rpm.

His pelvis's average rotation is 100 rpm.

The average force his front leg comes down is 175 percent of his body weight.

Leon Halip / Getty Images

Chapter **2**
"I Think" Teaching vs. Science Teaching

For years, since professional baseball started in the 1870s, pitching coaches have been preaching to their students: "that's just the way it's always been done." That was the way they had been taught—and if it worked for them, it would work for the latest youngsters too. Coaches have been teaching young pitchers how to achieve the positions of the body in the same way that their predecessors did 100 years ago, and they believe them all to be correct. This is what renowned baseball instructor Dick Mills calls "base-belief" coaching. Maybe it's due in part to baseball's nature as a tradition-rich sport, but pitching has often been taught according to history. In fact, since Cy Young started throwing overhand late in the nineteenth century, little has changed in pitching philosophy—until recently, when science started to creep into the thought process.

In recent years scientific approaches have made inroads, just as the statistics boys have made inroads in baseball's front offices. Still, according to Dr. James Andrews's research, 70 percent of current major league pitchers have poor mechanics.

> **According to Dr. Andrews, 70 percent of current major league pitchers have poor mechanics**

On the other end of the spectrum from base-belief coaching is evidence-based coaching. Based on biomechanics and medical orthopedics, evidence-based coaching uses the latest technology and tools available. This approach acknowledges that scientific studies and video analysis have determined what happens in the pitching motion and how it *should* happen for optimal results and reduced injury.

Evidence-based coaching aims to get the pitcher into certain positions that take advantage of the body's ability to produce the kinetic changes that translate into energy, and to move that energy up through the body and out through the shoulder. All of the positions presented in *The Physics of Pitching* are based on scientific case studies that determined the optimal positions that are needed to take advantage of the kinetics of proper biomechanics—for the body to create kinetic energy and move it along naturally. The scientists set out to prevent injury, but through video analysis they actually discovered what made all great pitchers great. If it was indeed a Gordian knot, they untied it, with studies of the hardest throwing pitchers in baseball throughout their careers—pitchers such as Sandy Koufax, Bob Feller, Bob Gibson, and Nolan Ryan, all the greats for whom they could find moving pictures, analyzed at 100 frames a second.

The studies found that all these great pitchers had the same thing in common. They all end up in the same position, in the

Cy Young was one of the first pitchers to achieve great success throwing the ball overhand. The award for the best pitcher in each league is named for him after a career that lasted from 1890 to 1911, during which he won 511 games—a mark unlikely to ever be surpassed. *Prints and Photographs Division, Library of Congress*

same common ways—a long stride, a closed position, superior hip and trunk rotation, good arm extension, weight transfers, and weight shifts; they all had good acceleration and deceleration. The studies looked at all the things that go into how the body can throw 100 miles per hour, or 95 miles an hour. All the pitchers who throw with a high velocity have remarkably similar attributes as they get ready to deliver the ball; however, they get there from their wildly varying windups.

But there some coaches teaching out there who really don't have any idea where the energy comes from. They teach what they think it's supposed to be—not what it actually is. You could call that "old school," for lack of a better term.

For instance, years ago they used to teach that you have to "drive off the rubber" or "dip and drive." Most pitchers can't "dip

Sandy Koufax, one of the most dominant pitchers in baseball history, won three Cy Young Awards (at a time when only one award was given for both leagues) and threw four no-hitters and one perfect game during his short career. He retired at the peak of his success, at age 30, due to arthritis in his elbow. Notice that his pitching hand is facing toward third base. Koufax made that adjustment so quickly it's difficult to see without video. *Focus on Sport / Getty Images*

Very few pitchers possess the combination of strength and skill that Tom Seaver did. This photograph demonstrates the power required to succeed as a "dip-and-drive" pitcher.
Focus on Sport / Getty Images

and drive" and be successful. You have to be a tremendous, strong athletic specimen to be able to dip your body and drive it out over 100 or 120 pitches a game, without getting tired. So you have to be very passionate about strengthening your body and staying that way. You can't be weak and succeed as a "dip and drive" pitcher, like Tom Seaver was. He was an unusual physical specimen, and he took great care of his body in order to be able to drive off the mound 150 pitches a game. But not everybody comes with that package, or that determination and passion to work on his body.

The other thing biomechanic scientists found through video analysis was that with "dipping and driving," the upper body had a tendency to get ahead of the lower body. The back foot was pushing the upper body out before the lower body was ready, which caused a lot of arms to drag. If the arm drags, you're just throwing with your arm; you've lost all the energy that was created by the legs and the trunk rotation. This makes pitchers slower, and their pitches flatter.

Mike Marshall's Approach

Former Cy Young Award winner and iconoclast Dr. Mike Marshall has developed several advanced training programs for pitchers. His 280-day program is designed to provide young pitchers with an injury-free motion and better velocity. Marshall, who pitched in a major league record 106 games as a reliever with the Dodgers in 1974, has been shunned by Major League Baseball, perhaps in part because he refuses to change or compromise any part of his complex system, which takes coaches and trainers at least two years to learn to administer.

In baseball, time is money, and few teams are willing to experiment in new techniques for young pitchers who show great promise and have already proven they can pitch. Nobody wants to take a kid who throws 90 miles per hour and enroll him in a rigid 280-day program to learn a "healthier" way to do it. They're afraid he'll lose that high-end velocity, which is already rare enough.

These teams may want to reconsider. Take Mark Prior, for example. He looked unhittable when he came up as a young pitcher with the Chicago Cubs, but some scouts predicted a short career because of the way he threw the ball. Sadly, they were correct, and his major league career was over by the time he was 25. This is exactly the kind of outcome that Marshall strives to eliminate through his program.

Focus on Sport / Getty Images

When you "dip," you also pound about two-and-a-half times your body weight into the ground over the rubber, and there is no value in that. Instead, you should be getting explosively into the power slot position. (We discuss getting into the positions and achieving them in Chapter 5).

There is some confusion, though, because you'll hear coaches who really know what they're doing say, "Don't push off the rubber." What they really mean is that you don't want to *consciously* drive off the rubber. If you're consciously driving, your tendency is to get your upper body over your lower body. So just have the back leg and rear hip get the body going into the power slot

Nolan Ryan offers a perfect example of what the power slot should look like. His back leg is totally extended; his front leg is slightly flexed but firmly braced. His foot is flat on the ground and slightly turned in. His chest is directly facing third base. Notice how the front shoulder is closed, relative to the plant foot. His shoulders are level and his head is straight. This is why Nolan Ryan threw extremely hard.

position, not so much the back foot, that *consciousness* of the foot. Starting forward this way will get you into the power slot explosively and keep your head centered on the triangle formed by your legs as you land in the power slot position, and not tilted forward. Then you are able to get that great velocity out of the hip and trunk rotation.

Another "innovation," if you will, is the long stride (though a number of greats used one). A lot of guys out there used to teach shoulder-width strides, just pretty much short, comfortable strides. But studies have found that a long stride creates energy from both the large and small muscles of the leg. The muscles stretch, the muscles contract, the front foot lands on the ground and that kinetic energy shoots up the pelvis and the trunk, creating more momentum. Then, as the body rotates, the rotational forces create more velocity.

Another older but still prevalent idea is the value of long distance running for pitchers. There is a lot of literature through the years about how pitching coaches were of different minds about the amount of running their charges did, with some quite fanatical about pitchers strengthening their legs by running "poles," which is from one foul line to the other (and back again) at the edge of the outfield.

Strength and conditioning specialist Eric Cressey, who contributes in Chapter 14, is no fan of endurance running for pitchers. He points to a study of 16 college baseball players and programs that was done for exercise professionals who wanted to develop evidence-based training programs. The study evaluated the compatibility of cardiovascular endurance and neuromuscular power training. Sixteen Division I collegiate baseball players were divided into two training groups, with their lower body power measured before and after the season. The first group performed moderate- to high-intensity cardiovascular endurance training three to four days a week throughout the season, mostly running. The second group, on the other hand, participated in speed/speed endurance training, including many sprints of varying lengths. The study found a significant difference between the groups even during the baseball season. The scientists examined the data and concluded that, for baseball players, athletes who rely heavily on power and speed, conventional baseball conditioning that involved significant amounts of long-distance running should be altered to include more speed/power interval training—in short, more sprinting, less jogging.

In addition, as Cressey details in two web postings he cites in Chapter 14, long-distance running can have quite a few more negative consequences that could inhibit the development of

> **Studies have found that a long stride creates energy from both the large and small muscles of the leg**

velocity and make it harder for a pitcher to stay healthy. Still, to this day, major league teams have guys do 30 and 40 poles, running miles and miles. They're still teaching what they think it's supposed to be. A lot of college and amateur programs have their pitchers run and run and run some more, when they really should be doing strength training and sprint work. Pitching is strength, agility, and speed, and the exercises have to address those muscles and systems.

Brian Pendergast shows a good example of the "long stride," which studies have shown to be a crucial element in maximizing pitch velocity.
Courtesy Fordham University Athletics

Chapter **3**
There Is
No Easy Path

"Throw like Bob Gibson," the pitching coach teaches—except that most people can't naturally rotate their trunk as quickly as Bob Gibson did, and no coach can change that.

So advice like that isn't much good to anybody but Bob Gibson, and he's retired. Gibson generated so much energy in his trunk rotation that he could throw the ball 90-plus miles per hour, and that plus curve certainly helped his fastball. But he was also fearless on the mound, and that is something you can certainly learn from Gibby.

Bob Gibson did not take his craft and ability for granted. There is no secret to success on the mound, there is only hard work and dedication to your craft. You can't take it lightly—work on it one day, leave it for another. In addition to tempo and rhythm, a good pitcher needs stamina and focus. And the stamina and focus carries over off the diamond; you have to work at it every day so it becomes as natural as tying your shoelaces.

Learning proper mechanics can add velocity, stamina, and control, provided the pitcher exercises the proper commitment to the endeavor. No amount of "knowing how to do it" lasts, if that is all you bring to the table. This is true for every age group. Even if you believe in your superior talent, it must always be combined with a great effort. A lack of effort will get you nowhere in this game.

Superior talent must always be combined with a great effort

Hall of Famer Bob Gibson threw with such force that the momentum created by his tremendous hip and trunk rotation left his body facing first base after delivering the pitch. Pitchers should ideally finish in a square position, facing home plate—but Gibby was the exception to many rules.

Establishing a Routine

How do you work? Everybody's routine is different, but I'll give you my routine. Of course, what works for me doesn't necessarily work for somebody else.

Let's say I was a starter in a five-man rotation; I start today, the next day I'm off. On my off days, I need to be working on my throwing program, throwing 15 to 18 minutes to keep my arm strong. Long toss: first 40 feet, then 60, then 90, then 150, and on back. Younger kids should work up to only 120, 130 feet, they should never continue if they're struggling to throw. Then I would throw a sideline, my bullpen, the next day would be throwing program, and then I'd pitch in the game the next day, every fifth day.

You play catch every day. You have to because you have to train your arm to be able to hold up. When you get to the Major League level, you're playing 162 games, and if your arm is not in shape, you're going to fall out halfway through the season. It's true for younger players too; your season is just as rigorous for you (except for all those rainouts). The old saying is, "It's not how you start, it's how you finish."

Rusty Meacham

Be a coachable youth—listen; if you subscribe to his ideas and he is able to understand what you want, do what your coach says. Otherwise, say, "No, thank you. It's not helping me." We're all human. Nobody is going to hurt you.

Find a coach who teaches evidence-based coaching.

At the same time, think of it as a journey. If you are a freshman who doesn't make varsity, it won't be the first time a deserving player is overlooked—and it won't be the last. Don't expect a rosy outcome every time. It wouldn't be much of a journey if it wasn't interrupted by some pitfalls.

The 18-minute Throwing Program

Younger kids—start playing catch at 40 feet, then go to 60 feet apart, then 90, then 120, and maybe 130.

Older kids—40, 60, 90, 150, and on back as far as you can without struggling to throw.

Then, you work back in on flat ground, and you get your throwing partner down into the catcher's crouch, and you work on getting your pitches down in the zone. Then, it's on to working on your pickoff moves and things like that.

That's the limit of your off-game pitching. A pitcher can't pitch all day, but a hitter can hit all day. Now, that's an exaggeration. A hitter can't hit all day, but he can hit a lot longer than we can throw.

When I was working out of the bullpen, to warm up, my routine was to go five pitches away to a right-handed hitter if I'm a right-hander, then five in. Obviously, if you're a lefty, it's the opposite. Then I would work on my off-speed stuff. That gets me ready for the game.

When you're on your non-pitching days, if it's your bullpen and you're throwing a full side, you don't want to throw more than 40 or 45 pitches. Usually you're 5 out, 5 in, then you work on what you had trouble with the last time you pitched. Even if you pitched a perfect game, there was something there you could have done a little better. Always strive to get better, even when everybody tells you you're hot stuff.

Then when you're throwing a light side, you should limit it to 15–20 pitches. A light side would be for a guy who wants to get a little touch-and-feel workout. For instance, if a guy is going to start on a Saturday, he might throw 15–20 pitches in a light side Friday, just to get the edge off.

Of course, everybody has a different routine. What worked for me might not work for somebody else. Some guys throw 10 in and 10 away, some just 2. Whatever works for you. But you have to have a routine because you can't go down to the bullpen and just throw with no idea in your head. A routine will help you concentrate on the job at hand, not what you want to throw on the fly. How you throw in that bullpen is going to carry right on over into the game.

You'll certainly run into that on the mound. The ball scoots through the second baseman's legs and you have to get four outs instead of three. Get used to it. Shortstop throws the next one away and now it's five outs, and it seems like you have to strike everybody out if you want to get out of the inning. Don't yield to it. Keep pitching your game.

Getting frustrated leads to walks or leaving the ball up in the zone. Neither one of which is going to make the situation any better. Bear down and get the next guy. It's like a bad golf shot: Forget it and move on to the next one and make it a beauty—because you can.

It Takes Passion—And Work

One of the things I find interesting is that most young pitchers I work with today, whether it be in Little League or Senior League, high school, and even college, have a dream of making it in the big leagues and they think their coach can make it happen. They've all had this dream in the back of their heads ever since they were little kids. They want to step on the big field. They want to play in "The Show." So these kids—and now I'm talking about any sport—hire pitching coaches, gymnastics coaches, figure skating coaches, golf coaches, whatever it may be, and expect these guys to make them an Olympian, a major league pitcher, a Division I pitcher or hitter, or a professional golfer. It's true that a good coach can help, but it's not the big answer.

It's about the individual having a passion to be really, truly the best he can be, and that takes work—a lot of work.

I always give my daughter, Heather, as an example.

At six years old, she began to figure skate recreationally, and by the time she was eight years old she was very good at it. She was so good at eight years old that a number of high-profile coaches across the country thought she should work with them, so she could reach her goals and dreams. As young as she was, she knew what she wanted to be—she wanted to be an Olympian.

When your kids are that young, you have to ask, "How do you really know you want to be an Olympian?" How do you really develop passion at that age? Of course, as adults we may forget how much passion an eight-year-old is capable of. We may only think of lapses of concentration, some things we see as reckless behavior. Of course, some kids have the passion to work that hard for the chance to be an Olympian. Heather did.

It means those athletes, gymnasts and figure skaters, especially, work five to seven hours a day, seven days a week, trying to master their craft at a very young age. To reach this goal is such a long shot, but these youths have the willingness and the desire to do it.

I find it interesting that when I meet some of my students for the very first time, I always ask them, "Outside of your regular practices, how much do you work on your craft all by yourself? How many hours do you work a day on your craft?" And I get the same answer. It doesn't matter what level, Little League to college. There's almost a hesitation. "Not that much. Maybe once a week. Maybe throwing a ball," a student says.

My whole point to this is that I have these young people who come to me, that reach out, that want me to take them where they want to go, but they don't realize I'm not the magic bullet, that I'm not the answer to it all. It really requires them to work very hard every day, during the season and out of the season.

Len Solesky

If you truly want to be a Division I pitcher or a major leaguer, that's what it's going to take. Not once a week, not three times a week. Every day.

I've just always found it funny that all these high-profile athletes with big dreams want to just basically go about it on their natural ability. If you really want to get there, though, it takes work, day in and day out.

You think you have it tough as a pitcher learning your craft? Let me tell you about hard work. I remember when my daughter, Heather, had learned all her single jumps. Then she had to learn her double jumps. One of the biggest hurdles to overcome for all figure skaters is executing the double Axel, which you need to do in order to compete at a high level. For most skaters, it's not something that happens easily. It usually takes them 8 to 12 months to learn that one particular jump.

Len explains to Connor how to point his off arm elbow at the target, keeping it at shoulder level. This should be done as the plant foot hits the ground, keeping the front shoulder closed and allowing the upper and lower body to rotate explosively to the plate. If the front shoulder is slightly open, all the energy that you have built up will fire off early and you're just throwing with your arm.

You are taking off from a small, little, forward outside edge on your left foot and landing on a small, little, back outside edge on your right foot, doing two-and-a-half revolutions in the air, with a 38-inch vertical leap. It really has to work precisely, and technically perfectly—or you fall or stumble. There is no room for error. The technique has to be perfect, which is why people cheer when they see it done.

Then the mental part comes into play once you have proven to yourself that you have the technique. Now you have to have the confidence to do it over and over again, flawlessly. There will always be doubt. Doubt is the great destroyer of athletes' goals.

Yet, a lot of young baseball players today—when they start working on a particular pitch or their swing in hitting and they run into a little difficulty—tend to give up on it so easily. It's because they want it to happen so fast. They just want to get it so quick. For example, the curveball: instead of learning the technique properly, they just want to get it and throw it, as opposed to truly mastering its rotation.

Figure skaters spend almost a year perfecting that double Axel and they work at it seven days a week, four or five hours a day. And they fall on their backsides four or five hours a day, seven

Michael demonstrates how to break the hands down and up, and get the throwing arm into a high cocked position, with the off arm elbow level with the shoulder.

Len demonstrates with Rusty Meacham that when landing in the power slot, you need a flexed front knee, but firmly braced back knee, to balance and stabilize. This position will allow the upper body to finish up over the lower body in the deceleration phase. If you land on a straight stiff leg, your upper body is going to have a tough time finishing off over the lower body as you release the ball.

days a week, for nearly a whole year, before they get the jump the first time. They're all waiting for that jump to work right the first time, to get it all working together. The whole point is that they don't give up on it. They know they have to have it to compete. There is no getting away with not having it. It's incredible the black and blue bruises and injuries they bear in trying to achieve that one jump.

Then once they've got that jump, figure skaters have to learn the triple jump—that's three-and-a-half spins in the air. Again, they have to stick to it.

Young pitchers today may be working on their changeup, and if it is not going easily, their tendency is to stop working on it. And they never develop it. Then they wonder when they get to college why they don't have it. They didn't put their work into it. They didn't stick to it. They didn't play catch with it for just 15 minutes a day, to master the feel and grip of that pitch. The same thing with the curveball, the same thing with all their pitches, the same thing with their location. They didn't master it; they didn't

Austin Young demonstrates a great finish position. His head is tracking the target, and his hips and shoulders are square. His right heel is slightly toward third base. Austin is 13 years old and has worked very hard over the past two years on all his positions.

stay focused on it. Pitching is about target practice. If you don't work on hitting your location spots all the time, and settle for throwing the ball down the middle of the plate, you're not going to succeed at the next level.

The moral of the story is that it's easy to work on what comes easily to you. It's not so easy to work on what comes slowly and difficultly. But you will enjoy the rewards—a curveball that breaks sharply; hitters swinging and missing at changeups that end up out of the strike zone; and, especially, that back-up slider that cuts the black for a called strike three.

I'll tell you another interesting story that reflects the thinking in parents' minds sometimes. This is an over-the-top example, but it speaks to the way some parents think, and most parents think at least a little bit.

My daughter was training out in Colorado Springs at the Olympic Training Center with the famous and renowned coach Carlo Fassi. He was the trainer and coach of Peggy Fleming, Dorothy Hamel, Robin Cousins, and many other great Olympic skaters.

I can remember Carlo and his wife, Christa, taking me out to lunch when I was visiting my daughter for a short weekend. We got into a conversation about my daughter's progression and so forth. And I'll never forget Carlo turning around and saying to me in his Italian accent, "You know what, Len, it was so funny. I have this very wealthy Texan, oil tycoon, whose daughter I teach figure skating. He said to me, 'Look, I want my daughter to be an Olympic gold medalist. I will pay you two million dollars to make that happen.' "

Fassi said he told the wealthy Texan, "I would love to take your two million, but there is no way I can guarantee you that I can make your daughter an Olympic gold medalist, because I have no idea what her work ethic is or whether she has the passion to reach that goal. It takes a tremendous amount of both, and it also requires an awful lot of talent."

Yes, you have to have talent as well. Nobody is going to take a below-average athlete and turn them into what in their parents' mind is the best, a high-level athlete.

Even if someone turned around and offered me $2 million, there's no way I could guarantee a kid could play in the major leagues. It takes a lot of work, and the players still have to have talent. If you put it all together, well, that's how dreams come true. The reality of it is that there are no guarantees, so you better have a passion for your life goals.

Now, my daughter became a pretty high-profile junior skater, then a senior skater. No, she didn't make the Olympic team. All the way through that journey, she worked and lived away from home starting at age 10—2,200 miles away in Colorado Springs. She went to Italy, Greece, Cape Cod, spent all her young life away from home. She didn't reach her dreams, but she became pretty good and could have skated professionally. So if you really want something, you have to be passionate about it; you have to want it, and you have to work hard. You have to have that desire constantly. And you have to have a plan. Part of the plan I propose to my students is, "Your family comes first, school comes second, your sport comes third, and all the fun things are all on the bottom." All those things need to be balanced in the plan.

Chapter 4
Harnessing Kinetic Energy

Today, doctors, physical therapists, athletic trainers, researchers, and academics all study kinesiology, the science of human movement. The term comes from the ancient Greeks' word for movement: *kinesis*.

The kinetic chain involved in pitching encompasses a coordinated human movement in which both energy and momentum are transferred up through body segments to achieve maximum magnitude in the final segment.

During pitching, the shoulder exceeds 7,000 degrees per second of internal rotation for adult pitchers. It is considered the fastest human movement, a movement generated by the windup process. The concept of a kinetic chain is developed from the idea that the energy expended in the pitching process is created with large muscle segments and is transferred through the legs and trunk, out to the throwing arm, wrist, and then eventually the ball. For example, the kinetic chain for throwing consists of the legs, hip, trunk, upper arm, forearm, hand, and the baseball. Pitching's kinetic chain includes a sequence of motions: the stride, rotation of the pelvis or trunk, upper torso rotation, elbow extension, internal shoulder rotation, and wrist flexion.

The potential velocity at the distal end where the ball is released is greater if more body segments contribute to the total overall force. Less energy is required if the kinetic chain is executed properly, if the pitcher's mechanics are correct. And the

> **The performance of a pitch is greatest when the kinetic chain is unbroken**

performance of the pitch—whether its velocity, its movement, or its location—will be improved when the chain is unbroken. But if "the gate breaks," one part of the body gets ahead of another, or goes off center, all three key priorities for success are lost and can't be regained on that particular pitch.

Tim Lincecum of the San Francisco Giants is a prime example of a pitcher whose style uses these principles of kinetic movement. A small guy, especially by major league standards, he was downgraded

Tim Lincecum, who learned how to pitch from his father, wasn't picked until the forty-eighth round of the major league draft, by the Chicago Cubs in June 2003. He decided to go to college instead. After four great seasons at the University of Washington, he was picked tenth overall by the Giants in 2006. *Courtesy Ronald C. Modra*

by many scouts who doubted his durability, despite his collegiate star status in the Pacific-10 Conference for two years running. Now, he's a Cy Young Award winner, because despite his small frame, he throws the ball past more hitters than just about anyone else in the National League and is routinely at the top of the majors in strikeouts. How does he do it? An aggressive, long foot plant allows him to maximize his hip rotation, while delaying his back shoulder and arm as long as possible. This high-velocity hip rotation, generated by his legs, along with his loose and delayed torso pulls his upper body along at high speed, delivering his arm at the same time.

The same principles are at work with the great pitchers of the past. If you watch old video of Bob Gibson, you'll see a phenomenally quick and powerful hip and trunk rotation. His style was different,

Bob "Rapid Robert" Feller went straight from the family farm in Van Meter, Iowa, to the big leagues at the age of 17. His family built a nonfictional "field of dreams" on the farm, and his father coached and recruited the Oakviews to play on it. *Courtesy Bob Feller Museum, Van Meter, Iowa*

but the kinetic mechanics worked the same for Gibson; the big muscles in his legs led to a strong trunk rotation, which enabled him to deliver the ball with great speed and accuracy.

Similarly, video of Cleveland Indian great Bob Feller, who threw over 100 miles per hour in an age when that was considered undreamed of, shows a very unusual style. But as Bullet Bob gets into his power slot position, all the mechanical parts have come together to generate the kinetic power that produced 3 no-hitters, 12 one-hitters, and over 2,500 strikeouts in an 18-year career, interrupted by four years in the navy during World War II. Feller also had tremendous trunk and pelvis rotation.

Interestingly, different pitches seem to produce some changes in mechanics, despite a pitcher's best efforts to deliver the ball exactly the same way. According to studies in the *American Journal of Sports Medicine* and the *Journal of Applied Biomechanics* (conducted by Dr. James Andrews and his colleague Glenn Fleisig, among others), fastballs had significantly greater pelvis (600 degrees per second) and upper trunk rotation velocities (1,120 degrees per second) than curveballs (560 and 1,070 degrees per second) and changeups (540 and 1,020 degrees per second). The curveball had significantly more forearm supination, when the palm is turned upward (32 degrees), than the fastball (17 degrees) or changeup (18 degrees).

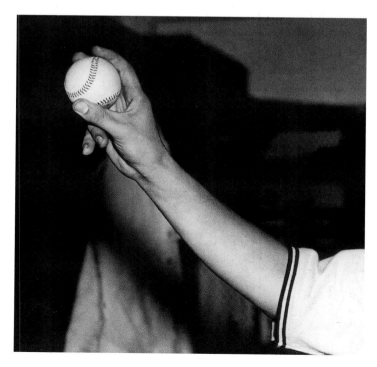

Bob Feller demonstrates his curveball grip. *MVP Books Collection*

Who Was the Fastest Ever?

We've talked about a lot of the greats in this book. Everybody has heard of Bob Gibson, Bob Feller, and Sandy Koufax. But who was the fastest pitcher to come down the pike? Would you believe "Nuke" Laloosh, the character played by Tim Robbins in the movie *Bull Durham*? Well, his real-life counterpart anyway, as encountered by the writer/director of the 1988 film, Ron Shelton, who was once a player in the Baltimore Orioles organization. Shelton based his iconic character, in part, on Steve Dalkowski, who played nine years in the minors before his arm gave out.

Shelton wrote about his inspiration in the *Los Angeles Times* in 2009, calling him "a little guy, which was shocking at first, with short arms, thick glasses and an easy smile." He was 5-foot-11 and 175 pounds, according to Baseball-Reference.com, which is small for any professional athlete. Other sources say he was 5-foot-8 and "chunky."

Dalkowski was born in 1939 in New Britain, Connecticut, and was signed out of high school by the Orioles and sent to Class D ball (which might be the equivalent of a rookie league today). His minor league statistics are sketchy and a little thin until 1962. Baseball-Reference.com shows

Tim Robbins as Nuke Laloosh in *Bull Durham. Orion Pictures/ Photofest*

that Dalkowski played that year in the Class A Eastern League in Elmira, New York (192 strikeouts to 114 walks in 160 innings, win–loss record of 7–10, with a very respectable 3.04 earned-run average). He probably should have washed out immediately because of his wildness, but the lure of that un-hittable speed in his fastball was irresistible—so the Orioles stuck with him.

Longtime Orioles manager and premier umpire-baiter Earl Weaver took a particular interest in the prospect. Weaver saw all the great fastball pitchers, including Koufax, Sam McDowell (who was also a tad wild), Dick Raditz, Nolan Ryan, Gibson, and he said, "Dalko threw harder than them all."

According to the *Sporting News*, Weaver said, "He was unbelievable. He threw a lot faster than Ryan. It's hard to believe, but he did."

Steve Dalkowski, pictured here during spring training with the Baltimore Orioles, was the inspiration for the character "Nuke Laloosh." *National Baseball Hall of Fame*

Longtime baseball field and general manager Gene Mauch listed Dalkowski as number 5 in his top 10 hardest throwers when asked by the *Sporting News*.

Because the radar gun was not then the extension of a scout's hand that it is today, Dalkowski's true measure was never ascertained, but many thought he could throw in the neighborhood of 110 miles per hour. His wildness was the stuff of legend, and it kept him out of Major League Baseball. In 995 minor league innings, he walked 1,354 batters and struck out 1,396. In a Double A Eastern League game, Dalkowski struck out 27 and walked 16 while throwing 283 pitches. In another game, he didn't finish the second inning because he had already thrown 120 pitches. In high school, it was said, he had pitched a no-hitter while walking 18 and striking out 18.

Walter "The Big Train" Johnson played his entire 21-season career with the Washington Senators. Johnson was one of the fastest pitchers of his era, and pitched his way to a record 110 shutouts. *Prints and Photographs Division, Library of Congress*

So many legends have grown up around Dalkowski that it's tempting to think of him as a modern-day Paul Bunyan. Among the stories are the following:

- Ted Williams stood at the plate one time in spring training against Dalkowski and said he was the "fastest I ever saw" and wanted no part of the pitcher after that.

- Teammate Herman Starrette once bet Dalkowski he couldn't throw a baseball through a wall. Dalkowski set himself up 15 feet away from the wooden outfield fence and hurled the pill right through the boards.

- He once threw a ball at least 450 feet on another bet. He was supposed to throw the ball from the outfield wall to home plate, but he threw it well beyond the plate, over the 40-foot-high backstop, into the press box.

- In 1960 at Class A Stockton, Dalkowski threw a pitch that bypassed his catcher and broke an umpire's mask in three places, knocking the umpire 18 feet back and sending him to the hospital with a concussion.

- He was said to have taken off part of a hitter's ear with an inside pitch.

In his first two seasons of professional baseball in the Appalachian and South Atlantic Leagues, he averaged 19 strikeouts and 18 walks per nine innings.

But according to Shelton, "Though he terrified hitters, he rarely hit a batter. Cal Ripken Sr., his catcher through much of the minor leagues (and one of my managers) said, 'Dalko was the easiest pitcher I ever caught. He was only wild high and low, rarely inside or out—but the batters didn't know that.'"

Still, the Orioles and Weaver, then a manager in the organization's farm system, wouldn't give up. Shelton says Weaver convinced Dalkowski to take a little off the fastball. By 1963, he had begun to throw more strikes than balls. In one stretch for Elmira in 1962, Weaver said Dalkowski had a 52-inning stretch where he struck out 104, walked only 11, and allowed 1 earned run.

So, in spring training in 1964, Dalkowski was poised to join the major league club. But as he was fielding a sacrifice bunt by Yankees pitcher Jim Bouton (later the author of *Ball Four*, the book that gave commissioner Bowie Kuhn ulcers with its tales of the hijinks of major league icons), Dalkowski felt a pop in his arm. He never regained his fastball. He was finally released after the 1965 season, his potential, like that of so many young hopefuls, dashed by injury.

A troubled man, Dalkowski then drifted, drinking heavily and toiling as a migrant farm worker in California. He was arrested many times for fighting and other minor offenses.

In the end, though, he was rescued, first by his wife and, after she died, by his family in Connecticut, where he now lives in a home for the disabled. He was honored in Los Angeles in 2009 when he was inducted into its Baseball Reliquary's Shrine of the Eternals along with Roger Maris. He got out of his wheelchair to throw the first pitch at a Dodgers game.

Now the nuts and bolts. The idea is to get the pitcher into certain positions that take the best advantage of the body's ability to produce the kinetic changes that translate into energy, and to move that energy up through the body and out through the shoulder.

All of the positions presented in *The Physics of Pitching* are based on scientific case studies that determined the optimal positions that are needed to take advantage of the kinetics of proper biomechanics for the body to create kinetic energy and move it along naturally.

Chapter 5

The Seven Key Components of the Delivery

Despite his mechanical quirks, Fernando Valenzuela always seemed to get the job done. This book will show you a more conventional path to successful pitching. *Getty Images*

If you watch many baseball games on television or listen on the radio, you will hear a lot about a pitcher's mechanics. There are a lot of clichés in baseball, but a color guy talking about a struggling pitcher's difficulty with his mechanics has got to be right up there with "who's on first." That's because mechanics are the key to pitching. It's a reasonably unnatural act, pitching a baseball, but believe it or not there's still a science to it.

Now, some guys can get away with some crazy things—Jonathan Papelbon comes to immediate mind or Fernando Valenzuela for you old-timers (and the young guys, too, thanks to the movie *Bull Durham*). But the fact is that in the end they always manage to end up in the right place when they deliver the ball. They were so good that nobody was going to change what they do. We're assuming you want an easier path, so we'll bypass those approaches. Instead, this is a simple way to achieve proper and effective mechanics that will work for the young aspiring pitcher.

The objectives of developing good mechanics are basically four in number. The first objective is an important one, to prevent injury; the second is to increase velocity; the third, to increase control; and the fourth, to improve consistency in your command of your pitches—to get your off-speed stuff to work, and your breaking pitches to break the right way.

In the end, you want to get your body in the right position so that the body can deliver the arm and the arm can deliver the ball. You want to do that with good tempo and good rhythm. You want to start right and you want to finish right. A good delivery and good mechanics are not easy to come by. You can have good mechanics on one pitch, and three pitches go by and

Think Buddha and Keep Your Head Straight

As this book covers the basic parts of the delivery, I want to give you a very simple rule. It's a little Zen-like.

In all the phases and all the mechanical positions that a pitcher has to achieve properly for his delivery, his head should be lined up directly over his belly button. Your belly button is the center of gravity of your body—it's the center point; it's the balance point. So when you start in the stance, your head is in a direct line over your belly button. When you rotate and pivot and get into the balance position, your head stays over your belly button. When you land in the power slot position, your head is in a direct line over your belly button. When you get to a square position after you rotate to throw the ball, your head is not tilted off to the side, is it? No, it's over the belly button. And then when you finish, your head is in a direct line over the center of the belly. That's your center of gravity all the way through. If you can do that in all the facets of your delivery, you'll have good mechanics.

Practice in front of a mirror and get the idea in your head so that you become one with the middle of your body. I told you it's a little Zen-like. But pitching takes a lot of mind control. You have to practice with your mind as well as your body, but we'll get to into that later in the book.

your mechanics can fall apart. That's not good mechanics. Having good mechanics is a consistent delivery of every pitch. It's a very tough thing to do; that's why mental toughness is such a key part of baseball, particularly for pitchers. Pitchers start every play in baseball, and they need a laser-like focus on each pitch.

The steps introduced here allow the body to work together mechanically, like the cogs and gears in a fine watch, creating and transferring kinetic energy in a consistent fashion. We're going to show you not just how to get in the positions, but why you get into each position and what you accomplish with this technique. That ends up being the motivation, because you see what the value is: to maximize your velocity, to maximize your command, to maximize your control, to maximize your movement. That's what it's all about. And most important—injury prevention, to take the stress off your arm. You should keep as much stress off your arm as you possibly can.

We will utilize a long stride, getting all the body parts in the right position at the right time to deliver the arm.

With the long stride, when the legs stretch, the muscles contract. When that front foot hits the ground in the power slot position, the kinetic energy shoots up as the hips and pelvis rotate which in turn creates more energy and more speed as the trunk continues rotating the body 180 degrees. That energy then transfers to the shoulder and comes out the arm. There should be little real effort on the arm, because it becomes a slingshot in effect. You are throwing effortlessly, with your entire body. Simple.

If it's effortless, then it shouldn't hurt you, either. Provided, of course, you do not overuse the arm, pitch in too many leagues, or for too many months. Forgive us if we say that more than once in this book, but it's the biggest cause of injury among young pitchers. Many kids end up with dead arms in high school and never get the chance they could have had to play in college. Bear in mind, though, that pitching overhand is an unnatural thing to do and the possibility of injury is always present. You can do everything right and still hurt your arm. Good mechanics or bad mechanics, every time you throw a fastball you risk injury.

Before getting into the five basic components of the delivery, it's important to stress that it all depends on balance, tempo, and rhythm. Those three things are the key to keeping your mechanics in sync. If you can work on getting your body in the right position with balance, tempo, and rhythm, you will have better command and control, which leads to success.

Balance is everything. As soon as you tip out of balance, the mechanism that is your body delivering the pitch falls apart. One part can't keep up with the rest and several things happen. Your arm can't catch up to the rest of the body. It basically drags, resulting usually in a high and away pitch, or too far in. You lose velocity and risk injury.

Now let's talk about the basic positions.

When the pitcher first addresses the batter from the stance, the glove should be square to the target and positioned just below or even with the chin.

1. The Stance

The body should be square to the target. The pitcher should be in a tall, athletic position. The glove should be square to the target and positioned just below or even with the chin. If this isn't comfortable for you, find a location between the belt buckle and chin that suits you, and keep the glove in line with the belly button, which is your center of gravity. The pitcher's feet should be positioned with the balls of the feet resting on the front of the rubber. The right-handed pitcher begins with his weight on his right foot. The feet should be shoulder-width apart. When the pitcher begins his delivery, he shifts his weight to his left foot. This begins the rocker.

Though this position seems very simple, it's critical to the rhythm and momentum of your delivery. Think about it this way: if you don't take proper advantage of the starting position, you might as well just throw from the stretch. That's how important it is. After all, there is a 1-2 mph difference in velocity between the stretch and the stance. Can you afford to leave that speed on the table? There is a big difference between a 78 mph pitch and one that comes in at 80 mph.

This is a fine place to talk about mirror-work. One of the most common faults Len finds in his students is the inability to recognizing weight shifts. They get on the rubber, they square up, they take a step either back or to the side, and they don't take enough time to experience the weight transfer. They go right into their pivot and lift their knee up. They assume they are getting into the balance position, never actually feeling the transfer of weight. You should constantly work to master your balance over the rubber, working on each of the different weight transfers—from the rocker, to the balance position, and then the weight transfer of the lower body in the power slot position—all with tempo and rhythm. The best place to do this is in front of a full-length mirror.

Brian Pendergast demonstrates the stance.

The pitcher should be in a tall, athletic position. He should be focused on his catcher, relaxed, and balanced.

Standing tall, the pitcher's athletic posture should create a vertical line between the tip of his chin and his feet.

The head is centered over the belly, the body's core.

The glove should be square to the target and positioned just below or even with the chin.

The body should be square to the target. Weight is evenly distributed.

The pitcher's feet should be poised with the balls of the feet resting on the front of the rubber. His weight should be balanced.

2. The Rocker

From the stance, the pitcher takes a short step straight back, approximately four to six inches, transferring his weight back to his left foot (for a right-hander; of course, it's the opposite for a left-hander). The pitcher should feel like he is rocking. While transferring his weight back, the pitcher should keep his head over his posting foot, which is the leg he will balance on (right for right-handers, left for southpaws). His head should never get behind his posting or pivot foot. This will cause the pitcher to lose balance.

The rocker is a transfer of weight from right to left to right. It creates momentum and energy, and begins to transfer energy from your legs through your, hips, trunk, and all the way out your arm. You can actually feel it when you find that rhythm, and it should carry you all the way through the delivery.

When you work in front of the mirror, always work at game speed, not 60 or 70 percent. You want to emulate the game situation, to recreate the adrenaline rush of throwing the first pitch of a game. If you do mirror work at anything less than game speed you won't gain the muscle memory you are going to need for game situations. If you practice swinging a golf club at 50 percent, can you expect to get anything out of it? Or, let's say, a boxer works at 50 percent attacking a heavy bag, would he ever develop a punch at that pace? No, and neither will you. Mirror work is a great way to see evidence that you're completing each position and weight transfer, from beginning to end. Remember, the mirror doesn't lie.

The rocker is the first weight transfer that begins the kinetic energy train.

There should be very little movement of the head or glove as the pitcher focuses on the catcher's target.

While transferring his weight back, the pitcher should keep his head over his posting foot.

Head is still and in line with the center of the body.

From the stance, the pitcher begins his delivery by taking a short step straight back, approximately four to six inches, transferring his weight back to his left foot. He should feel like he is rocking.

3. The Balance Position

Once the pitcher has completed his rocker or weight transfer, right foot to left foot (again, the reverse for a lefty), he should then transfer his weight back to his right foot. This is accomplished by lifting the heel of his right foot and rotating on the ball of his foot to a parallel position to the rubber. This is the pivot. The pitcher should still be in the tall position. At this point, the pitcher should simply lift his knee up—not kick it, jerk it, or swing it—to at least a waist-high position. The higher you can lift, provided you can balance yourself, the greater force you will generate when you explode forward. The more force, the more energy you can generate in your legs, the greater your velocity.

When lifting the knee, keep the toe pointed downward. The front foot should be directly under the knee, not outside the knee. If the foot is outside the knee, the tendency is to swing the leg, possibly causing the pitcher to throw across his body, over-rotate, and throw horizontally to the plate. When the pitcher throws across his body, his shoulders rotate too early and the throwing arm drags behind and can't catch up.

Keep the foot directly in a straight line underneath the kneecap. The pitcher's shoulders should be level. The front shoulder should be aligned straight to the target.

The front hip bone should be aligned straight to the target, and slightly ahead of the shoulder. The posting leg should be straight and the knee should be slightly flexed, but firmly braced. The head should be directly over the posting foot. The weight should be on the ball of the posting foot and flat. The glove should be held chest-high and straight over the belly button.

It makes no sense to continue the windup if you don't achieve this position. If the balance position isn't achieved, the upper body will start to go forward and the lower body will stay behind. It's critical that you feel your weight shifting back to the posting foot when you lift your leg up.

In the balance position, the pitcher should be in a tall position. We don't want the upper body leaning over the lower body—straight means straight, standing tall.

The eyes are tracking the target at all times, especially after the release of the ball. Mentally lock in at this point and stay locked in all the way through.

The front foot should be directly under the knee, not outside the knee. The front toe should be pointing downward. This will allow the pitcher to lift his knee higher and will also help him land on the ball of the foot. It's not just an aesthetic thing.

The posting foot should be parallel to the rubber, with weight bearing on the ball of the foot, not on the heel. You won't be able to balance on one leg on your heel. You'll tip over. It's that simple.

The shoulders should be level and in a straight line directly to the plate. Don't coil or turn the shoulders away from the target. The front shoulder should act like a gun sight; it should never turn away from the plate.

The front hip should be in a direct line to the target. The front knee should be at least waist-high, or 90 degrees minimum. Remember: the higher a pitcher loads—with balance—the stronger he will unload, resulting in a faster pitch.

The posting leg should be straight, and the knee should be slightly flexed and firmly braced to balance and stabilize the body. If the knee totally straightens, the pitcher will be out of balance.

4. Hand Separation

Once the balance position is achieved, the body is now ready to move explosively to the target, eventually into the power slot position. All the body parts must work together as a well-tuned mechanism. When the knee reaches its highest apex in the balance position, the back leg and hip drive the body forward explosively; the knee drops and the hands separate simultaneously downward. The thumbs of the hands break down over the belly button. The hands then come back up so that they are at 90-degree angles to the body.

As soon as your lower body starts to move from the balance position, your hands must move. They move in coordination with the lower body. Some kids aren't able to identify the exact point at which their knee reaches its highest place in the balance position. They think they're there, but they may only be halfway, so they start going out early. As long as the hands move with the lower body, you will reduce arm drag and maintain the timing of your mechanism.

This begins the trunk rotation.

The back leg and rear hip get the pitcher started forward. The front foot is still up off the ground, toe pointing downward, in a straight line with the plate. The back foot and hip now drive the lower body explosively forward, leading out with the front hip and the front leg.

The hands break at the same time the pitcher begins to drive to the plate, shifting his weight from his back leg forward, using his back hip to get started.

Hands break in the center of the body.

As the hands break out of the glove, they move down, back, and up like a pendulum swinging. The hands start close to the body and the elbows lift them away.

The front foot leads the knee, the knee leads the hip, and the hip leads the shoulder, moving sideways toward the target.

Eyes are focused on the target. Head is level and in line with the belly button.

Back is straight.

To keep arms aligned with the trunk and the plate, the glove elbow is used as a gun sight to the plate. Lead with the elbow and the glove will follow.

Fingers must be on top of the ball as it comes out of the glove.

The front toe points down.

5. The Power Slot Position

As the leg and hip now drive the lower body explosively forward, the front hip and leg lead out. The front foot stays parallel to the back foot and turns upon landing. As the lower body is driving forward, you should feel your back leg extend. As the front foot extends and is ready to touch down, the back leg is fully extended. The throwing arm should be in a high cocked position, and the off arm (glove-arm) elbow should point toward the plate. The thumb of the glove hand should be pointing down, which allows you to remain in a closed position. The elbows are shoulder height, at about a 90-degree angle; the ball in the pitcher's hand should be pointed toward second base, the fingers should be on top of the ball. The shoulders and chest should be in a straight, aligned position with the target.

When landing on the plant foot, the pitcher should land softly on a flat foot, with the toe turned slightly inward at about 12 to 15 degrees. The weight should be on the ball of the foot. The front knee should be flexed and firmly braced at about 25 degrees. The body should be facing third base.

When you land on that front foot, you have about 150 percent of your body weight landing. Just like a jet plane, if you land on it too hard you're going to crash. Your mechanics will fall apart, and you will be throwing with nothing but your arm.

The pitcher's eyes are tracking the target, and his head is directly over his belly button in a straight line.

The body should be in a closed position when the plant foot hits the ground, and the back foot is still in contact with the rubber, allowing the back leg to fully extend.

The lower body should be facing third base as well. The knee should be slightly flexed and firmly braced.

The front foot should be slightly closed upon landing, 12 to 15 degrees. The foot should also be flat on the ground with the ball of the foot bearing the weight.

The throwing hand should be hat-high, facing between the shortstop and second base. The elbow should be slightly flexed, mirroring the front elbow in alignment.

The shoulders are level in a straight line to the target. There should be a minimum of 90 degrees under each arm. The thumb in the glove should be facing straight down, which will keep the front shoulder closed before the body begins to rotate.

The trunk of the body should be facing third base.

The front elbow should be pointing toward the target in a straight line.

Rear leg should be fully extended and not bent.

When the front foot plants, the rear foot should still be in contact with the rubber, ready to turn and rotate.

6. The Square Position

With the front foot firmly planted and braced, the pitcher squares his chest to the target at home plate. The pitcher achieves this by rotating his hips and trunk and turning on the ball of his back foot. His chest and his body are square to the target – this is critical, because you want all your energy going toward the target.

The pitcher's glove hand should be ahead, even with the shoulder, with the elbow gliding inside toward the hip, not outside the body like a chicken wing. His pitching elbow will be slightly above his shoulder, his hand outside his elbow but slightly flexed. The hand should be on top of the ball.

Resist the urge to consciously pull the glove hand to the chest. Let the chest come to the glove naturally. Many kids are taught to pull in that glove hand in an attempt to get more energy into the hip and shoulder rotation. Video analysis suggests that this tends to make the shoulder dive forward too quickly, not allowing the hips and trunk to rotate with the upper body. The upper body gets ahead of the lower body. The shoulders over-rotate, not allowing the arm to catch up. Pitchers who pull the glove hand back tend to un-square their shoulders, so that the release point is not directly in front of them, as it should be, but early and high. The arm drags because it can't catch up to extend and release the ball out front.

After this position is achieved, you want to get the arm fully extended out in front of the body and to release the ball in front of your face.

With the front foot firmly planted and braced, the pitcher squares his chest to the target at home plate.

The throwing arm should be parallel to the side of the head. The elbow should be at least shoulder-high. The hand should be slightly outside the elbow, not inside.

The back foot should be parallel to the front foot, and the toe should be in contact with the ground, balancing and stabilizing the back side of the body.

The hand should be square to the target and the fingers should be on top of the ball. The wrist should not be bent. The hand should be straight up from the wrist.

Again, the shoulders are square and level to the target. The eyes are tracking the target. The head is directly over the belly button.

The glove hand should be out front and even with the shoulder, and should glide inside the body, not outside.

The hips should be square to the target.

The front knee should be firmly braced and flexed somewhere between 22 and 90 degrees.

The landing foot is still flat on the ground, toe turned in a minimum of 10 to 12 degrees, and the ball of the foot still bearing the weight. The foot becomes a claw to pull the body forward explosively.

7. The Finish Position

With the front knee braced and the front foot flat, the back leg drives the back hip into rotation around the front leg. The trunk rotates, generating energy and pulling it up from the legs. The arm will naturally come forward, delivered by the rotating trunk. The head stays above the center of the body. As the hips come square to the plate, the back foot releases and balances the finish.

As the arm fully extends, the head is still level and tracking the target. The front leg is still flexed and not stiff. As the ball is delivered out front, the arm continues and ends up outside the landing leg. Always continue your follow through. If you don't, you'll lose velocity, just like you lose distance on a golf swing when you don't follow all the way through.

The upper body finishes over the flexed and firmly-braced knee of the front leg. The throwing shoulder should be buried toward the ground, so that the hitter sees the back of your pitching shoulder. Your back should be flat, parallel to the ground.

At this point, your hips and your shoulders should be square to the plate, putting you in a good fielding position

As you finish your pitch, think of your arm as a car, smoothly decelerating about 10 mph at a time—100, 90, 80, and so forth. At this pace, the driver is in full control of the car. On the other hand, if you slam on the brakes at 100 mph, that control is completely lost. The same is true as the body propels your arm through to the finish position. An abrupt finish will negatively affect your control.

A proper finish reduces stress on the arm and also allows the arm to continue on its path of acceleration to maximize velocity.

The pitcher's throwing hand should finish outside his left thigh.

The pitcher's feet should be close to parallel to each other.

The eyes are tracking the target.

Chin is at the glove. After he finishes, the pitcher should be able to balance a book on his head.

From the catcher's view, shoulders and hips should be square to the target.

A good finish puts the pitcher in a good defensive posture, instead of falling off to the first base side.

The Stretch

Pitching from the stretch, when you have runners on base, is really not much different than pitching from the windup in terms of mechanics, except that you lose the momentum and energy you get from the additional weight transfers at the beginning of the windup. The only difference is the setup. You still go to the balance position, to the power slot, to the square position, and to the finish in the same way, with the same mechanics.

In the stretch position, the first thing you want to do is set up your feet. Your back foot should be parallel to the rubber. The back knee should slightly flexed but firmly braced. Your front foot should be parallel to the back foot. They should be about shoulder-width apart.

Your front foot should be ahead of your knee. Your knee should be ahead of your hip, and your hip should be ahead of your front

He's a closer by definition, but the Yankees don't hesitate to bring Mariano Rivera into the game with runners on base in the eighth inning. Why? He has nerves of steel and pitches very well from the stretch. *Getty Images*

Your setup should be the only element of your mechanics affected by the stretch. Every other position should be achieved exactly as when pitching from the wind-up.

shoulder. This will ensure the lower body will go out ahead of the upper body on its way to the power slot position.

The shoulders should be straight to the target.

The glove should be in the center of the body, chest to belly button.

The head should be directly over the posting foot, the back foot.

The elbows should be relaxed into the body.

A right-hander's chest should be facing directly toward third base, or first base for a left-hander.

Once the delivery starts, when lifting the knee into the balance position, the weight is shifted back to the posting foot. Once the weight shift is completed, the lower body weight is transferred, leading out with the front hip to the target, separating the hands simultaneously. The back leg and the rear hip should drive the lower body to the power slot.

Chapter **6**
Rusty Meacham on the Pitches

To best visualize these pitches, it's a good idea to take a baseball in your hand. Things will become much clearer in your mind with the object in your hand.

The Two-Seam Fastball

For the two-seam fastball, I would hold the ball between the seams with the two fingers, the index finger and the middle finger, close together, and I would go more toward the top of the ball, the slippery part of the ball. Everybody holds the two-seamer a little differently. Some hold the ball on the seams; I found it worked better for me to be more on top of the ball. Your hand naturally pronates (flexes forward) on the fastball, and I would have it come off my index finger and get that nice down and in action to a righty.

I really liked to use that pitch when I wanted to get a ground ball, moving it down and in on a righty and sinking the ball down and away to a lefty. I didn't like to come with the two-seamer as much when I was 0–2, because it moves so much and I didn't want to take the chance of hitting the batter in that favorable count. I'd run the four-seam in there more. You don't want to hit him when you're up 0–2.

The Four-Seam Fastball

The four-seam fastball you hold across the seams. I like to hold my fingers close together like I do on the two-seamer, with the index finger close to the middle finger. The reason is that the farther you have your fingers apart, the more you're going to slow the ball down; it gets to be more like a changeup or a split-finger. I used the four-seamer because it was a straighter fastball and I could throw it harder. I could come in on rightys or lefties 0–2 and not risk hitting them. It had more velocity on it, and it was straighter.

Whether I threw it for a strike or not depended. A lot of times, I would throw the four-seamer if I was ahead in the count to kind of move them off the plate, or maybe throw a four-seamer away, or maybe double up back inside. It all depends.

Colorado Rockies phenom Ubaldo Jiménez has one of the best four-seam fastballs in the big leagues. The aptly nicknamed "U-Ball" is frequently clocked in the 100-mph range. *Getty Images*

The Fork Ball / The Split-Finger Fastball

There's a difference between the fork ball and the split-finger fastball. I threw the fork ball with the seams, but I would jam it back in my hand. I would split my index finger and middle finger, and my thumb just kind of rode along the bottom of the ball. Everybody throws their balls differently, but I would turn my pitch over like my fastball. And I would get that nice 12–6 action. It would look like a fastball, and then when it got to you, it would drop off the table. I would throw that pitch basically down the middle. If I throw a split-finger or a fork ball and it's a called strike, then I'm not throwing it right. Am I going to complain if it gets called for a strike? No, but that's not my plan when I'm trying to throw this pitch. I am trying to throw this pitch so that it looks like a strike, and you have to chase it because it's in the zone. And that's how most pitchers get outs anyway, by throwing balls. They appear to be a strike, but if you let them go they'd be a ball.

The difference between a split-finger and a fork ball is the split-finger is not as spread out. It's going to have a sharper and harder break down. Mine was a little slower. They're both changeups, though.

Some guys throw both a changeup and a splitter. Some guys throw both a slider and a curveball. It's very tough to throw both a slider and a curveball. Most of the time, a pitcher will throw one or the other.

Dan Haren has a feared split-finger fastball, coaxing batters to swing just as the ball drops out of reach. *Getty Images*

The Changeup

I had what's called an OK changeup, or a circle change. I held mine basically like you would say OK to somebody with your fingers. I would grip the ball with the horseshoe always pointing to the right, and I would grip it with my three fingers, my middle finger, my ring finger, and my pinkie finger. I would pronate my hand when I would come out front and have it come off those three fingers to get that nice downward action away from a lefty hitter. I didn't throw it to a right-handed hitter. I liked it more for left-handers, because I used my slider and my split-finger/fork ball more for the right-handed hitter.

Len Solesky on the "Box" Changeup

Let me explain the box changeup. The middle finger is centered between the two seams; the index finger and the thumb are on the side of the ball. The little finger and the ring finger are on the other side of the ball. Face the ball toward you, and you will see that the fingers form a box. The ball is held somewhat back in the hand. When throwing this pitch, the tension of the grip should be as though you are going to throw an egg or a tomato—or a jelly doughnut.

The point is: Don't squeeze the ball, because the tighter you squeeze the ball, the harder the pitch will be, and you'll have less control over it. When throwing this pitch, make sure that you throw it with the same arm speed as your fastball. Trust that the grip will take enough velocity off the pitch to make it effective. After all, the most deceptive part of the pitch in the hitter's eyes is your arm action. When throwing the pitch, pronate your wrist and hand inward on release of the ball. This will give the pitch a down-and-away movement.

This is a great changeup for younger pitchers because their hands are small, and it is hard for them to grip a circle change.

The Slider

The slider can be held many different ways. On the baseball, you have a horseshoe; I like to hold it with the horseshoe pointing down, with my index and middle finger on the right part of the seam. And I would come through when I'd get out in front with a couple of fastballs, and then I'd just do a slight tip of the wrist, making sure my hand stayed on top of the ball where the ball came in looking like a fastball. All the pressure is on the index and the middle finger, and when you release it you press on the pad of the middle finger and turn it over. Then, right when it got to the hitter, it dove down maybe an inch or a half an inch, but it looked like my fastball—with a nice tight rotation to it.

Randy Johnson was one of the most intimidating pitchers ever to take the mound. "The Big Unit" consistently threw a 100 mph fastball, but his wicked slider was even more impressive. This pitch, dubbed "Mr. Snappy" by Johnson himself, disguised itself as a fastball and broke sharply at the last moment. *NY Daily News via Getty Images*

The Curveball

I wasn't a big curveball guy, but to throw a curveball, I threw it pretty much the same way as the slider, but put it back in my hand more. Again, I got my hand on top of the baseball, to get that good 12–6 (like on the hands of a clock) break to it. First of all, there a number of different grips to throw a 12–6 curveball. Usually, I position my hands as if you're looking at a four-seam fastball.

More Thoughts on the Curveball

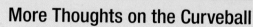

I put my middle finger inside the seam. Then I put my thumb behind the seam underneath the ball, at a 7 o'clock position (if you hold the ball out in front of you, you would see the middle finger at a 1 o'clock position and the thumb at 7 o'clock). The index finger lies on the skin of the ball next to the middle finger. It's very important when you are throwing this pitch to stay on top of the ball.

When throwing this pitch, you want to be thinking fastball, fastball, fastball arm speed. Our objective is to get the arm extended and get the ball in front of the face. When the hand is in front of the face, pull the hand straight down in front of the chest, almost as if you are pulling a window shade. When pulling the ball down in front of the chest, you want to flex the wrist forward, rolling the middle finger over the front of the ball and the thumb rolling over the front of the ball, spinning the ball in a 12 o'clock-to-6 o'clock direction. After releasing the ball, you want to immediately wrap your arm around your waist. Your palm should be facing upward.

The Knuckleball

The knuckleball is thrown, understandably, with the knuckles. It's the method of Red Sox pitcher Tim Wakefield, the most successful knuckleball thrower in the major leagues. Wakefield was arguably going nowhere as a position player in the Pittsburgh Pirates organization until someone saw him throwing the knuckleball in an everyday catch-and-throw at camp and suggested he give pitching a shot. The rest is history.

The single most important thing about the knuckleball is to throw it for strikes, which is tough to do with a pitch you basically can't control. Throw it for strikes, or guys will just watch it go by. You need them to swing at it. If they do, they're likely to pop it up or hit fly ball outs. It's a great pitch. You don't throw it hard, so you can come back more quickly and pitch another day. I've toyed with the knuckleball as a path back into the major leagues at age 42.

There are several ways to grip the knuckleball. As shown in the close-up photo, you can grip

Tim Wakefield throws a fastball that rarely exceeds 75 mph, but that doesn't seem to matter much. Wakefield learned to throw the knuckleball from two of its masters, brothers Joe and Phil Niekro. Wakefield's famous pitch dances, flutters, darts, and dives its way past hitters' bats—and sometimes past his catcher! *Getty Images*

it with two knuckles against the seams, the thumb under the ball, and the other two fingers on either side of the ball to help direct its path. Tim Wakefield takes a slightly different approach, with both the ring and pinky fingers on one side of the ball and the thumb on the lower portion of the other side. I throw it with four knuckles on the seams.

The knuckleball's premise is that it dances on the wind. Without spin, it won't fly straight, which is the intention. A windy day is perfect for the knuckleball—then it dances all over the place.

The intention is to throw a ball that doesn't spin with the stitches, and so does not follow a predictable path. You push the ball out, instead of throwing it. The pitch is very easy on the arm, with minimal stress put on the shoulder's labrum or on the elbow. Follow your usual motion, and then just push the ball toward the plate. You'll be a great knuckle-baller if your pitch is a strike more often than not. Even when it's a strike, it's hard to hit, hard to square the ball up, and its lack of velocity makes it harder to drive. But of course, there will be days when the pitch is flat, and they'll cream it. Such is a pitcher's lot.

Chapter 7
Common Faults and Corrections

In this chapter, we'll try to point out some common faults and corrective measures for your mechanics. Making a video of your efforts on the mound can help you identify some potential pitfalls. You should train yourself to make constant appraisals of your mechanics. Your coaches and parents can also spot these errors and help make the necessary adjustments to put you on the right track—throwing strikes, getting outs, and winning games.

Pitching instructor Bill Thurston notes that "most of the stresses that cause pitching arm injuries occur during the acceleration, release, and deceleration phases, but the stress may be caused by an improper technique that was used earlier in the motion."

In other words, if the body doesn't get in the right position at the right time, one segment of the body can create a cascading effect that causes injury. A sore foot can eventually hurt your shoulder, because, often unconsciously, you're making adjustments to compensate that will affect how the rest of the body works in the pitching motion. Similarly, one bad technique can throw the whole mechanism off.

The most common problem in pitching is **rushing the motion,** which basically means the upper body moves out ahead of the lower body. When you rush your motion, the body gets ahead of the arm, causing the arm to drag behind. Rushing causes the pitcher to lose all the energy from his legs. The solution: from the balance position, make sure you lead out with your front

One bad technique can throw the whole mechanism off

hip and your lower body. This should be part of your normal rhythm, and any deviation from that rhythm can lead to problems. A number of things can pull the pitcher out of rhythm, but the few most common are: A) he's trying to throw harder; B) he's nervous; or C) he is having trouble throwing strikes and begins to "aim" the ball. This all leads to one of the most common faults, **flying open.**

FAULT: FLAILING—The pitcher's arm is flailing behind his back. This is going to make it very difficult for the arm to get in a high, cocked position to rotate with the upper body when the front foot lands in the power slot position. His arm is dragging and will not be able to catch up to the shoulders as the body rotates.

REMEDY: STAY CLOSED—Colin gets in a good closed position in the power slot. He is landing in a straight line to the target. He is now ready to explosively rotate his pelvis and trunk to the target and a good finish position.

Fault: Flying Open

Opening early during the stride and cocking phase is common for pitchers of all ages. It's part of learning how to pitch. Young high school and college pitchers have to learn "stay closed" mechanics, and then, experienced pitchers must focus on "staying closed" until the stride foot plants. It has to be a mantra, repeated over and over in the pitcher's mind. When a pitcher is throwing high and away, (right to righty) or high and inside (righty to lefty), that's the giveaway that he is opening up too early. There are a number of other reasons for this, but the first thing to master is **"staying closed."**

Remedy: Stay Closed

Until the stride foot plants, giving the pitcher a stable base, the lead elbow, the shoulders, and throwing arm must stay pointed in a direct line to the plate. The head remains straight toward the plate, over the belly button.

This allows the trunk to transfer the power generated through the legs, as the trunk turns, to the upper body. The trunk's rotation must be in the proper rhythm and sequence (in relation to the other body parts in motion) to create the torque that is the major source of power and velocity.

The fault of "flying open" is that the upper body, shoulders, and arms open with the stride leg and front hip. This natural tendency decreases the force of the trunk's rotation and puts the motion out of sequence, weakening control and velocity, and at the same time, putting increased stress on the shoulder.

You end up throwing all arm, then. The arm will drag because the stronger and larger muscles of the legs, hips, and trunk have already fired and are not available to create torque. So all of your speed has to come from the shoulder and the arm, A pitcher throwing with just his arm wears out faster, not just in the particular game, but as time marches on, both seasonally and over the years.

Keep the front shoulder closed, the off arm out and elbow pointed toward the plate as you come out of the balance position. Be conscious of it as you first practice it until it becomes as natural as "flying open" was. Remember you have to learn how to pitch; you can't just do it any old way.

Now we'll look at some of the most basic faults in pitchers young and old. Remember that things can very suddenly get out of whack. A little adjustment can make all the difference. You will have to make adjustments during a game. After each game you pitch, you should review and then make adjustments in your next bullpen. Being aware of problems means you're halfway there to fixing them.

FAULT: ARM ALIGNMENT—Notice Brian's off arm is well below his front shoulder, which will cause his front shoulder to open up too quickly. On his throwing arm, his hand is too close to his head. He is over-flexed in his elbows. His hand is behind the ball.

Arm Faults

- **Long-arming, stiff-arming, the back swing** When getting into the power slot, instead of bringing the arm up into a high cocked position, the pitcher leaves it outstretched long behind him. It's tough for it to catch up to the rest of the body from back there. The pitch is likely to be high and away and the release point outside the body rather than directly in front. You lose energy and velocity with this cog in the machine lagging.

- **Flailing or stabbing** This fault is similar to long-arming, with the same consequences, but a little worse. Instead of bringing the arm up into a high cocked position, you let it drop and sort of flail behind the body. The flailing arm has trouble catching up to the body as the trunk rotates into the square position. It basically drags. This is usually caused by the shoulders moving out of alignment in the balance position, coiling your shoulders away when lifting the knee. You change the angles on your shoulders, so that when you separate the hands to go back and up, the pitching hand tends to get behind your back and flail.

FAULT: ARM ALIGNMENT—The pitcher's off-arm is way above his front shoulder, causing his back shoulder to drop down and no longer be level. His arm flails behind his back and goes long, making it hard to get into the cocked position in time with the pitching motion.

- **Wrist hooking** It looks just like it sounds. Guys who hook their wrist take the flexibility out of the arm, creating stiffness in the arm and the wrist. That can cause stress. A lot of guys start with their hand hooked over the ball, because all they ever heard was "keep your hands on top of the ball." So when they take their hands out of the glove, and break it down and up, they will either short-arm it (pull it out of the glove quickly and bring the arm straight back) or it almost makes you break your hand behind your bottom. Wrist-hooking makes it very difficult to throw a curveball, and when you throw a fastball, you can't get any pronation (forward flexion) of the wrist.

FAULT: ARM ALIGNMENT—The off arm of this youth league pitcher is very low and not level. His front shoulder has already opened up. His throwing arm is flailing behind his back, rather than in a high-cocked position. The front arm has already opened up and is swinging toward the plate. This early rotation won't let the arm catch up. The energy stays trapped in the lower body.

FAULT: FLYING OPEN—The pitcher has landed in the power slot and his front shoulder is wide open. The lower body is still facing third base and not rotating with the upper body. This is early rotation and the pitcher is throwing with just his arm.

- **Hand rolls under the ball** The fingers should stay on top of the ball, pointed in the direction of second base. Think again of being in a straight line to the target at the plate.

- **Overly short-arming** The arm swings back too close to the head into the cocked position. In other words, pitching as if you're a catcher. A great example of this is reliever Jason Mott of the St. Louis Cardinals, who throws in the high 90s and was a former catcher. When he gets wild, this is one of the reasons why.

Stride Faults

- **Direction problems** You can be too closed, landing in a diagonal from the plate. If your front hip locks (throwing across your body), you're too open and throw too short.

- **Landing on heel** If you land on your heel, a few things can go wrong. First, you can't balance and stabilize your body; second, your knee can't get flexed, which causes your leg to be straight; and third, your upper body won't be able to get over your lower body to finish the pitch or decelerate, causing stress.

- **Unstable landing** The leg and foot are not firmly planted and stable. If they are not stable, the body can't be stabilized. The body (trunk, hips, arm) can't be pulled forward.

FAULT: HEEL LANDING—The pitcher is landing on his heel in the power slot position, instead of a flat foot, which usually leads to a stiff front leg. The landing is unstable, and it's hard to flex the knee from this position.

- **Stride leg does not brace** When you land on your stride leg, the knee is slightly flexed. You have to brace upon it. You can't let the knee go forward; that will cause you to lose velocity because the body is going forward and not pulling through. Everything collapses. A long stride is good if the pitcher can get up over the braced stride leg since the ball comes out of the hand as it crosses over the stride foot. If you overstride, it's hard to get the upper body over the lower body as you finish the pitch.

FAULT: LANDING DIAGONALLY—Here's an example of a pitcher starting to rotate from the power slot. Notice his elbow is outside his left hip, almost like a chicken wing. This is going to cause him to open up too quickly and over-rotate his shoulders. As the ball is released his shoulders will be facing first base instead of directly toward the plate. He is landing diagonally. This will probably result in a pitch high and inside to a right-handed batter, with less velocity.

FAULT: UNSTABLE LANDING—This is a position you don't want to be getting into. This is getting into a square position. Notice David's front foot; he's on the edge of his foot and back on the heel. In this position, his foot can't stabilize his body, nor can he pull his body (trunk, hips, arm) forward. Notice the back foot hasn't even rotated, but is still parallel to the rubber, so the hips won't rotate, reducing hip speed and power. The elbow of his off arm is outside his body, instead of gliding inside his body.

Trunk and Upper Body Faults

CORRECT: LEAD WITH FRONT HIP—
Brian leads out explosively to the target
with his lower body and his front hip. This
allows the lower body and the upper body
to rotate together upon landing.

- **Rushed motion** The body is out ahead of the arm, with the usual bad consequences. The solution is to lead out with your front hip and your lower body from the balance position. Rushing typically occurs when a pitcher tries to turn too early, before the rest of the body is ready to deliver its energy.

FAULT: BODY ALIGNMENT— This youth league pitcher has tilted his trunk toward his back foot, throwing his body off alignment. He will get no benefit from the energy produced by his legs, which will force him to throw with all arm.

- **Upper body leads** Instead, the front hip should lead. If you don't lead out with your lower body into the pitch, the upper body gets ahead of the lower body, and, again, the swinging gate breaks and you lose power, control and velocity. An indication of this is when the pitcher seems to be tilting down.

- **Front side flies open before stride foot plants** You have to get in the power slot closed, but in a straight line. You have to plant your foot before you throw the ball.

- **Upper body flexes forward before the torso squares to the plate** Don't dive into the pitch.

- **Not being square** The shoulders and hips have to be squared to the plate.

Power Slot Faults

- **Elbow and hand too low** The elbow must be at a 90-degree angle to the shoulder.

- **Landing on a straight, stiff leg** The front knee is not flexed and firmly braced.

- **Elbow too high at stride foot plant** The trunk is forced forward, hindering the rotation of the hips and trunk.

- **Foot flying open** If you land in the power slot and your front foot is open instead of slightly turned in, it will cause the front hip to open up too quickly and open the upper body too quickly. You fly open when your front shoulder opens, and you can't throw a strike. The arm can't catch up, and whenever the arm

FAULT: STIFF LEG LANDING—This youth player's front leg lands straight, so he is struggling to have his upper body finish over his lower body. This is putting a lot of stress on his shoulder, not allowing the arm to properly decelerate.

FAULT: FOOT FLYING OPEN—An example of a pitcher turning his front foot over too quickly. This will cause the front hip to open up, and the back hip will turn along with it; the back knee will turn inward, and the back foot will peel off the rubber too quickly, causing the back leg to collapse.

FAULT: OFF ARM—The pitcher's off arm is rising too high, bringing the front shoulder up and dropping the back shoulder. Without level shoulders, it's almost inevitable that he will long-arm this pitch.

drags, you're throwing with just the arm, with none of the energy from the lower body transferring up. All your energy fires off early. This also changes the release point, so that you are not out in front of your body.

- **Early external rotation of the shoulder** If you cannot remain in a closed position, you'll open up, the most common error young pitchers make.

- **Improper off arm action and position, flying open early** The off arm is just a guide to get you to the slot in a closed position when you land. If you swing your arm out, you fly open.

- **Head not in the top center of the triangle** If the head is not over the center of the belly button, and tilted instead, you're done.

FAULT: ARM ALIGNMENT—During hand separation, when the hands break down, back, and then up, notice that the pitcher's palm is facing the shortstop side and not the first-base side. This is causing his arm to go long into the cocking phase of the throw. This will reduce velocity. Try turning your palm as you take the ball out of your glove and try to throw hard. You can't.

- **Wrist flexed versus extended back** Keep your hand on top of the ball.

- **Elbow flexed too much** Hand too close to head. Stay in your arm slot.

- **Palm faces forward too much** The ball should be facing toward second base.

- **Head, shoulders, and hip line not relatively level** Front side is elevated.

Acceleration and Release Phase Problems

- **Arm slot too high** Too close to the head.

- **Arm slot too wide** Too low; hand and fingers outside or under the ball.

- **Hand and elbow come forward in an upward plane.** Hand and elbow should come forward in a downward plane.

- **Trunk not squared to the plate** Upper torso flexes forward too early.

CORRECT: David Julian gets himself into a good fielding position after the release of the pitch. Notice that he decelerates his motion correctly, with his throwing hand landing on the outside of his posting leg.

CORRECT—Brian Pendergast is in a great square position, with his head over the center of his body, his leg is fully extended on the back side, his front foot is flat, balancing and stabilizing his body.

- **Elbow leads and continues forward too long on fastball** This causes the pitcher to look like a dart thrower.

Deceleration Faults
- **Head and shoulders do not come down over a braced stride leg** Hips stay behind the stride leg and the body recoils back.

- **Arm and hand cut across the body** You want a nice long arc of downward deceleration, outside the knee of the stride leg.

- **Arm extends outward, toward the plate** Remember, arm should decelerate in a downward plane. This is part of finishing the pitch.

Other Common Faults
- **The back leg collapses from the balance position** This causes the back knee to turn in an inward position and causes the front hip to open up too soon. It also keeps the legs from getting stretched out into the long stride. The solution is that once your weight is loaded over the back leg in the balance position,

CORRECT: LEG EXTENSION—Notice how Connor's back leg is extended, his front foot has made contact with the ground, his chest is facing the first-base side, and his off arm is in a closed position. What we want to focus on here is how Connor's back leg is extended; the knee is not bent and collapsed, or over-rotating at this point.

you want to drive off your back hip and leg and get the body moving sideways into the power slot.

- **Hands break too high** The hand goes straight back, instead of swinging down and back. You want to break the hands in the center of the body, between the chest and the belly button. The arm swings down like a pendulum.

- **The front foot turns too soon going into the power slot position** It needs to stay parallel and then turn over just before the front foot plants down. The solution is for the pitcher to stay sideways longer. That will give the pitcher a longer and more extended stride, and allows the legs to stretch.

- **The arm gets up in the cocked position too early** The solution would be to bring the arms down and up in a circle.

- **Ball has a slider-type spin** This a problem with younger kids and, if not corrected early, will stay with the kid until somebody shows him otherwise. It's basically throwing a fastball with your thumb on the side of the ball instead of underneath the ball.

FAULT: FLYING OPEN—Connor's upper body is in very good position, but notice how his front foot is slightly turned open. That toe should be in a closed position. By opening and turning that foot too soon, the front knee and front hip are starting to swing, causing the back hip to open up as well. The shoulders will rotate too early; his hips will rotate too early. His arm will try to catch up, but it will be kinetically impossible.

CORRECT: This youth league pitcher is looking very good as he lands in the power slot position. His shoulders are level as he prepares to bring his arm up into the cocked position. He remains in a closed position, and isn't opening up too early. The ball is still a little big for his hand, but his thumb is under the ball.

It kills velocity. With little kids, it may be because their hands are smaller and it's uncomfortable. But it has to be changed. It's uncomfortable for a while, but you lose anywhere from one to three miles an hour in velocity. You'll hear parents say, "Oh, my kid throws a natural curve." That's why. But what it can do is put a lot of stress on a young kid's elbow, because it's kind of liking throwing a mini-slider. It keeps the arm fairly straight, with no flex in the elbow, so you could end up with big elbow problems. The solution is simply to make sure your thumb is under the center of the ball. As small a problem as that is, it can be a big problem.

- **Change-up is too fast** This probably means the pitcher is gripping the ball too tight, creating more velocity. He needs to loosen up his grip, or tuck the ball back farther into the hand. You never want to slow your arm down. You want to throw with the same speed with every pitch. You throw the change-up just as hard as the fastball; it just comes out of your hand differently. That's what slows it down.

Chapter 8
Defending Your Position

Pitching is a lot more than just throwing the ball to the hitter. Being able to field your position ought to be very important to you. After all, superior fielding will get you outs that poor fielding pitchers miss. You don't ever want to give your opponent an extra out in a close game. That means more pitches for you, and it's demoralizing to your infield. Because you're pitching to contact, the sharper your infield, the fewer pitches you're going to need to throw in an inning.

Rusty Meacham says, "In the two years I was coaching with the Washington Nationals, we would sit in a room and we would go over our whole organization. If it comes down to me and you, and you field the position better than I do, and we both pitch the same way, who do you think is going to go? I'm out of a job. They're going to keep you because you're more valuable to the ball club. So it's not just throwing the ball and getting people out; it's what you do after the ball leaves your hand. Are you a good fielder? How do you handle yourself when you're on the mound? How do you handle yourself when things aren't going right? What's your demeanor? I've played with guys that hang their heads when somebody makes an error behind them. You can't do that. Pick your teammates up and they'll do the same for you when you're in a bind."

> You don't ever want to give your opponent an extra out in a close game

A key part in fielding your position is where you finish your delivery. After you have delivered the pitch, with your eyes never leaving the target, your pitching hand on the outer side of its

opposite thigh, you should have come down square to the plate in a fielding ready position. From here, you can grab intended bunts, swinging bunts (in other words, unintended), balls hit back at you, and balls tapped to either side of you. Finishing in a strong fielding position is a great form of self-protection, too. You're going to be as as close to the batter as anybody in the infield, and it pays to be able to field a screaming line drive hit back at you—those can really hurt or cause serious injury.

Be sure to practice pitchers' fielding drills as solemnly as you would a game, because those little bitty dinky hits will kill you. They tend to distract the pitcher afterwards, when they are not executed properly. The old "if only I would have . . . " bug gets you. Just remember, as Don Meredith, the great Dallas Cowboy

Rusty demonstrates hip and trunk rotation to the square position. A good square position keeps the body balanced and helps the pitcher prepare to act as a defender from a solid finish position.

quarterback and original Monday Night Football color man, used to say, "If 'ifs' and 'but, onlys' were candy, what a jolly Christmas we'd have." He's originally from Texas, if that explains it.

Another equally important part of fielding your position, if it's not pitching itself, is keeping the other team's running game—their ability to steal bases—in check. Taking the runner's legs from him, or at least the energy in them, by throwing over in a varying pattern helps your biggest partner in the game, your catcher. It also keeps runners from getting in scoring position and keeps double plays in order. All of which are most desirable. If you let them, other teams can run the life right out of your team.

Rusty gets himself into a good fielding position after the release of the pitch.

Rickey Henderson was one of the greatest base-stealers of all time. You can't always control the run game, but mastering your pickoff moves will help keep base-runners honest. *Courtesy Ronald C. Modra*

Pickoff Moves

Let's say you have a runner on first base. There are five ways a pitcher can control the running game. First, you throw over on the way up, when you begin to lift your leg in the balance position. The second look you give the base runner is when you throw over on the way down, as you begin to drop your knee. Three, you can hold the ball for three seconds or so and throw over. Number four, hold the ball for three seconds or so and step off the mound. Finally, the best one of them all, is to just hold the ball. When you hold the ball, one of two things will happen: The umpire will call time out or the hitter will.

Rusty Meacham says that's what he did with Rickey Henderson. "I would hold the ball on him. He was the toughest base runner I had to face. You vary your looks and never do the same thing. You don't want to get in a pattern the base runner can identify.

"Rickey, though, look at what he did. He stole, what, fourteen-hundred bases? I have had a lot of runners over there on base, but the toughest to keep from stealing was Rickey Henderson."

The idea is to stiffen up the runner's legs, even if you don't pick him off. Make the runner hesitate and you've won the battle. Every throw over should put a little doubt in the runner's mind, delay his jump, and give your catcher a better chance to throw him out. Have a quick pickoff move and a casual one. It's like changing speeds to a hitter. The slower move sets up the quick move. If you have a pattern, however, the runner is going to guess it right when he goes.

Left-handers, of course, have a distinct advantage with runners on first base. They're looking right at him, and they don't have to turn to throw over. As long as you don't land over the invisible 45-degree line, you can make the runner think you're going home. If he is stealing, and you fool him there, he should be out. If the runner is going when you throw over, though, make sure your first baseman has practiced that throw to second. He has to step forward to catch the ball and throw outside the path of the runner. Too many pickoff attempts end up with the runner safe at second base when the infield doesn't execute.

There is no excuse for a left-handed pitcher to lack a good pickoff move. Practice it until you get it right. That's one of the things you have to do in fall and early spring practices. Work on your pickoff moves.

This pitcher demonstrates a basic pickoff move to first base from a right-handed pitcher. There's nothing fancy about this move, and it's not going to produce many outs on its own—but it will keep the runner honest.

A pitcher has to work to control the running game. That also means you need to be quick to the plate once you start your delivery. In the big leagues, that's 1.35 seconds and lower from first movement to contact or when the ball hits the catcher's glove. "If your delivery is 1.4 seconds or higher, you're not giving your catcher any time to throw out the runner," Meacham says. "Obviously, 1.1 seconds is lightning speed. If you're at one-point-one, the only way a runner is going to steal is if he guesses right on you, and gets a good jump, but most of the time the runner is going to be thrown out by a catcher with a half-decent arm. If a catcher's time (from catching the ball until his throw reaches second) is 2 seconds flat and lower, he's pretty decent.

"Pudge" Rodriguez was 1.6 or 1.7. If you factor in a pitcher who is 1.35 or lower and a catcher who is 1.9 to 2.0 or better, then you have a pretty good combination. You can control the running game. But if your pitcher is slow and your catcher is 2.2, then you are going to have problems," Meacham adds.

Of course, younger kids can't be expected to hit those marks, and base stealers aren't going to be as good at getting jumps or as fast going to second base. We offer them more as illustration of what kind of time is necessary, and as a look into the big leagues. That's how good you have to be, and it doesn't hurt to have goals.

Now, let's look at some other pickoff situations.

Runners are at first and second. You have several options. First, let's take the **back-door pickoff**. You don't see this run often, but you come set looking at home. The first baseman is playing behind the runner. You look back at the runner at second and use your peripheral vision. When you see the first baseman break back to the bag, you spin and throw to first. With lefties, the first baseman has to key off their leg-kick.

Then, you have **timing plays**, where the pitcher is looking at home, looks back at second, and then looks home again. When the shortstop or second baseman, whoever has given you the sign, sees the back of your head, they recognize that as the key to break behind the runner to second base. Spin and throw the ball—like a quarterback and a wide receiver, it's about timing, and you have to practice to get right. The next play in your potential repertoire is the **"watch me" play**, where either the pitcher or the fielder

On a pickoff to second base, the pitcher spins quickly and fires to the bag. The play must be well-timed for the ball to meet second baseman or shortstop at the bag.

gives the sign (you can point to your eye, or something; though of course, you need to vary it). If either side doesn't answer the sign, the play is not on—you have to be on the same page.

"There has to be communication," Meacham says. "Then, you catch the ball back from the catcher, and you scan the infield. Say, for instance, I'm doing it with the shortstop. I look out there; he gives me the 'watch me' sign, and I answer back. I come up, looking at home. I always want to come up looking at home because if I looked in at second or first, I have messed the whole timing of the play up.

"So come up looking at home; look back at the shortstop if that's who you have the play on with. If he flashes you the glove, because he's got the glove on his left hand, that's your time to spin and throw. (A right-handed second baseman would flash you his throwing hand, and so forth). But you do not have to throw the ball."

The Slide Step

I am not a big fan of the slide step. With this move, the pitcher lifts his lead foot just a few inches, instead of lifting his knee fully into the balance position. Guys try to be too quick and they sacrifice their stuff to the plate. Why not be quick and throw a quality pitch to the plate, too? You can be just as fast lifting your leg up a little as you can slide stepping. I don't like slide stepping, because guys try to be too quick and they don't lift their leg up enough and their weight gets out too quick, and they have nothing behind it. It's kind of like getting a hitter on his front foot. Once you do that, what's he going to hurt you with?

As a pitcher, I want to get a hitter on his front foot, because then he's just arms; he doesn't have his full body behind his swing. It's the same thing with pitching: If you get out too quick, you don't have any power behind the ball. You don't have your lower half. When you slide step too quickly, all you have is your arm, no upper body. When you're throwing from the stretch, you simply can't afford to lose any additional power or velocity.

Next, there's the **inside move**.

"You come set and you lift your leg up. You want to sell it as if you're going to the plate. And you spin. You pivot on the ball of your right foot, and turn around, and throw the ball to second base," Meacham says. Again, you don't have to throw the ball. On the timing play, the "watch-me" play, and the inside move, you don't have to throw the ball. Be sure you have a good shot at getting the runner, because an errant throw that winds up in center field usually puts a runner at third. You don't have to throw when you make a move to second or third base.

Finally, there's the **first and third** situation. You'll see this play tried many times before you'll ever see it work. But sometimes it does, and that's an out you can really use. With runners on first and third, the pitcher fakes a pickoff to third, and then throws to first. Again, you don't have to throw the ball, either. It serves the same function as any move to first: It takes a little of the spring out the runners' legs. This move can also reveal whether a steal is on, and sometimes the batter may give away his intention to bunt.

Mark Buerhle is one of the best fielding pitchers in the big leagues. Winner of the Rawlings Gold Glove Award in 2009 and 2010, Buerhle makes some amazing defensive plays coming off the mound. *Courtesy Ronald C. Modra*

Andy Pettite's pickoff move is second to none, which makes it very difficult for base-runners to steal while he's on the mound. Left-handers like Pettitte have a real advantage here, as they're able to move toward first base with a quick, smooth move. *Getty Images*

Chapter 9

Recognizing and Preventing Injury

Common Causes of Pitching Arm Injuries

The most common causes of injuries to high school, college, and even professional pitchers are a lack of proper conditioning, improper throwing mechanics, and overuse or overload.

Longtime Amherst College coach Bill Thurston distinguishes between the two.

"Overload" makes a distinction between one event and overall overuse. An overload injury is the result of throwing too many pitches during one outing. Maximum pitch counts for various age groups, or for an individual pitcher's normal strength and stamina, are effective in preventing overload. The American Sports Medicine Institute has developed guidelines for different age groups.

Overuse, on the other hand, is the result of pitching too often without adequate recovery time or a good maintenance program between pitching outings. Of course, each individual pitcher varies in arm strength and susceptibility to arm fatigue, tightness, or soreness. So, each has different recovery needs.

Each individual pitcher varies in arm strength and susceptibility to arm fatigue, tightness, or soreness

For years, pitches have been counted and monitored for professional, collegiate, and high school pitchers in order to reduce the risk of injury. A 1996 survey conducted by the U.S.A. Baseball Medical & Safety Advisory Committee showed that most experts believed pitch counts should be kept for younger pitchers as well. The committee then sponsored an epidemiological study

All-American Brian Pendergast is getting into the slot position. Notice how his back leg is totally extended. You can actually draw a straight line from his rear hip to his ankle. This is great extension. Achieving each position with proper mechanics will decrease a pitcher's risk for injury.

by American Sports Medicine Institute (ASMI) to look at the issue. This study, published in the *American Journal of Sports Medicine* in 2002, showed a significant relationship between the number of pitches thrown and the risk of shoulder and elbow pain in youth baseball.

In fact, ASMI and the U.S.A. Baseball Medical & Safety Advisory Committee believe joint pain indicates the early development of a potentially serious joint injury. "If—for some valid reason—a league is unable or unwilling to enact pitch count limits, the league should limit the number of batters faced. Since nine- to twelve-year-old baseball pitchers average about five pitches per batter, pitch count recommendations can be converted into batter limitations by dividing by five," the institute and committee note. But since some young kids have more trouble with control than others, and so would throw more pitches in an inning, that seems a poor substitute. "Pitch limitations are a better choice than batter limitations for accurately monitoring and controlling risk of overuse," ASMI argues.

Pitch Count for Youth Pitchers

Many youth baseball leagues monitor and regulate pitch counts. If your league doesn't have such rules, make sure your child's coach is aware of your concern about his pitch count. American Sports Medicine Institute recommends these limits for youth pitchers:

9–10-year-old pitchers	11–12-year-old pitchers	13–14-year-old pitchers
50 pitches per game	75 pitches per game	75 pitches per game
75 pitches per week	100 pitches per week	125 pitches per week
1,000 pitches per season	1,000 pitches per season	1,000 pitches per season
2,000 pitches per year	3,000 pitches per year	3,000 pitches per year

According to ASMI, pitch count limits pertain to pitches thrown in games only. These limits do not include throws from other positions, instructional pitching during practice sessions, and throwing drills, which are important for the development of technique and strength. Backyard pitching practice after a pitched game is strongly discouraged.

Poor conditioning is another common cause of injury, and proper conditioning involves the entire body—the legs and core muscles, as well as the throwing arm. (Chapter 14 addresses strength and conditioning in greater detail.) Muscles also need to warm up before they are demanded to function at high levels. College-age players should have become accustomed to regimens already developed by the time they reach that point in their pitching lives. Younger players, however, may have yet to understand that they can't perform from a cold start—no more than an old car battery left outside in a Minnesota winter can. Coaches should supervise proper stretching and warm-up procedures before a ball is thrown. You begin throwing the ball slowly and closer to your partner, progressively increasing distances and intensity as your arm warms up. It's not a hard thing for any kid to understand if he's seen a ball game before it starts, or between innings, when the fielders practice as the pitcher takes warm-up tosses.

U.S.A. Baseball also notes that because a youth pitcher usually stays in the game at another position after pitching (as opposed to being replaced in the lineup), the player is eligible to return to the mound later in the game. But that is a bad practice in terms of the health and safety of the pitcher. Muscles need time to recover after they have been used, especially in the case of the strenuous activity of pitching.

Preseason is time for hope, but also can mean sore arms for pitchers who have not spent the off-season in a conditioning

An adequate warm-up routine with plenty of stretching will prepare your muscles to perform their best.

program. Those who haven't kept after it over the winter tend to throw too hard, too early. Keep a schedule of off-season throwing in the school gym. It all goes back to your work ethic and your passion to master your craft. Work on the things you don't do so well. For instance, maybe you've got a good fastball and you can spot it up pretty well. If your off-speed stuff is not as effective as it ought to be, you should focus on improving this aspect of your game. You have to have an off-speed pitch to keep hitters off-balance As important as balance is to a pitcher, it's equally as important for a batter—so you want to do whatever you can to disrupt the batter's balance. For younger pitchers, those under 16, a changeup is just as effective as a curve or slider, and it doesn't put as much stress on your shoulder and elbow, which haven't fully developed in most kids.

Today, more kids are training off-season, but you do need to take some time off from throwing in order to rest your arm and let it recover. That doesn't mean you shouldn't be working on your strength and conditioning, though. That's a year-round, day-in, day-out task, but it needn't be boring. Taking the tedium

Michael Pendergast finishes off a pitch. Notice how his back is flat. A smooth deceleration phase is essential to protect yourself from injury.

out of exercise is one of Eric Cressey's main concerns. His website (www.ericcressey.com) is a great resource, offering regular e-mails to address all kinds of issues for pitchers and athletes in general. Later in this book, he offers in-season and off-season training programs later to follow.

Another major factor in preseason injuries, Bill Thurston says, is that "pitchers are not working with a normal in-season rotation schedule and do not get enough recovery time from a lot of necessary drill work that involves throwing." He is referring to pitchers' fielding practice, which is held every day and for extended periods in spring and fall practices, and professionally during spring training.

Finally, as we've said in every chapter so far, improper throwing mechanics are a sure-fire path to injury. If a pitcher has bad mechanics anywhere in the delivery, there is a great chance of early fatigue, and fatigue just promotes even poorer mechanics, leading to injury.

Recognizing Pitching Arm Fatigue

Most kids want to compete and won't admit to being tired when a coach visits the mound. A minor pain isn't something they'll admit either, so parents and coaches have to be aware of signs a pitcher is tiring. And a pitcher can tire before his normal pitch count. Everybody has one of those days, right?

Here's what to look for:

- Is he starting to rush his motion, throwing high pitches? This is the most common symptom. He's rushing to get more out of the throw than what he has. But the body is tired and just can't do it. Even if he won't admit it to himself, his body knows. This is an example of how poor mechanics create even more stress because the arm now drags behind and can't catch up—the pitch becomes all arm. Of course, the pitcher is defeating his own purpose, because rushing tends to open the glove-hand shoulder, and so he will lose even more velocity When he starts walking people or leaving the ball up in the zone and getting hit, he is likely to be tired.
- Is he rubbing his arm between pitches?
- Does he start stretching his arm before he sets up to throw? He may shake it more than normally.
- Is he taking more time between pitches? Is he slowing down, walking around the mound more?
- Does he have redness or swelling in his elbow or bicep? This could also indicate some injury.
- Is he trying to overthrow the ball? This is akin to the first thing we mentioned that indicates fatigue.

Austin Young's front foot is firmly planted on the ground, balancing and stabilizing his body in the square position. Notice that his back foot is still in touch with the ground, which balances the back side of his body. Great balance sets pitchers up for health and success.

- The pitcher may shorten his arm deceleration path and not follow-through to the outside of his plant foot. He will lose his normal arm extension during the release and deceleration phases.
- If he is hurt, he will change his delivery and mechanics to protect his arm from further stress and pain. He probably isn't making this decision consciously. His body is making it for him.
- The pitcher may not get his hand and elbow up to the normal height in the cocked position. This is another mechanical flaw that puts more stress on the arm. It will appear that he is accelerating in more of an upward plane.
- Between innings, the pitcher may massage his elbow or top of the shoulder. With arm fatigue, a pitcher's hand can tremble.

Recognizing Injury

Every time you throw a fastball you risk injury. It's not a natural act, and a pitcher must repeat it over and over again. The muscle, tendon, or ligaments around the shoulder, elbow, forearm, and wrist are remarkably complicated mechanical wonders, but you're toying with their design when you pitch.

These signs should make you suspect injury:

- Redness, discoloration in the arm.
- Swelling, puffiness, stiffness, extreme tightness.
- A burning sensation upon movement.
- Sharp pain versus a normal dull ache.
- Lack of grip strength.
- Loss of extension, flexion, or normal range of motion.

If any of these signs are apparent, the pitcher should stop throwing, apply ice to the injured area, and take time off to

Mark Mulder joined forces with Barry Zito and Tim Hudson in Oakland, leading the A's to the playoffs in four consecutive seasons from 2000-2003. Here, Mulder's arm and shoulder are wrapped in ice after a game. Unfortunately, Mulder's career was cut short by a series of shoulder injuries. *MLB Photos via Getty Images*

recover. He should get a break from participating in drill work, throwing, and pitching assignments. The pitcher needs to keep up with his strength and conditioning work, although his throwing should be limited until he recovers.

If there is little improvement after rest, the player should be seen by a physician to learn the extent of the injury.

Showcases and Year-Round Play

Showcases offer young players the opportunity to display their skills to scouting services, college recruiters, and major league scouts. They tend to be pretty expensive, though, and there isn't much sense in going to one before you're a sophomore in high school. You can go as a sophomore to get on recruiters' radar screens, so that they'll be looking for you in your junior year, which is the critical year.

Recommendations to Avoid Injury

Based upon its expertise and review of existing studies, U.S.A. Baseball Medical & Safety Advisory Committee makes the following recommendations for minimizing a pitcher's risk of future serious arm injury and maximizing his chance of success:

- Coaches and parents should listen and react appropriately to a young pitcher when he complains about arm pain. A pitcher who complains or shows signs of arm pain during a game should be removed immediately from pitching. Parents should seek medical attention if pain is not relieved within four days or if the pain recurs immediately the next time the player pitches.

- The risk of throwing breaking pitches until physical maturity requires further research, but throwing curves and sliders, particularly with poor mechanics, appears to increase the risk of injury.

- Pitchers should develop proper mechanics as early as possible and include more year-round physical conditioning as their bodies develop.

- A pitcher should be prohibited from returning to the mound in a game once he has been removed as the pitcher.

- Baseball players—especially pitchers—are discouraged from participating in showcases due to the risk of injury. The importance of showcases should be de-emphasized, and at the least, pitchers should be permitted time to appropriately prepare.

- Baseball pitchers are discouraged from pitching for more than one team in a given season.

- Baseball pitchers should compete in baseball no more than nine months in any given year, as periodization is needed to give the pitcher's body time to rest and recover. For at least three months a year, a baseball pitcher should not play any baseball, participate in throwing drills, or participate in other stressful overhead activities (javelin throwing, football—as quarterback, softball, competitive swimming, etc.).

U.S.A. Baseball Medical & Safety Advisory Committee has reservations about showcases that occur near the end of the players' season, when players are often fatigued and need rest and recovery. "In other instances, players participate in a showcase after a prolonged period since their league ended and without adequate preparation to throw hard again. It is without a doubt that young throwers will try to overthrow at these events in an effort to impress the scouts and coaches, which further increases the risk of serious arm injury," the committee notes.

We tend to disagree a little in the case of showcases that are operated at a national level among Amateur Athletic Union (AAU) teams, because they offer a ball player an opportunity to play in front of scouting services and a host of recruiters and scouts from across the country; the more people who see you, the greater your opportunities. These showcases are part of your season and you don't go in cold, or after a period when you haven't been playing. If you are not playing on such an AAU team, and showcases are past your season, you should know you're going and stay in playing shape. Remember you're still working on your craft, even when you are no longer playing games. Just remember not to overthrow, which isn't going to help you anyway. (See Chapter 15 for more discussion of showcases.)

To get more opportunity to develop their skills and more chances to pitch, many young players play in more than one league. Although the amount of pitching in a league is often limited by league rules or the judgment of its coaches, nobody is checking individuals in more than one league (though, some coaches do pay attention to the issue and will check which players on their team pitch elsewhere and the last time they pitched). This is where that pitch count total becomes all the more important to keep track of—and days off after starts or long stints on the mound. Check the table earlier in this chapter for ASMI's recommendations for the number of pitches for each age group. The strength and skills needed to be a successful pitcher are developed by repetition; however, a pitcher must also give his body time to rest and recover.

U.S.A. Baseball medical group, with ASMI, also cautions against year-round baseball. "In certain parts of warm-weather states (Florida, Texas, California, etc.) baseball leagues are available in all seasons. However, the principle of periodization states that an athlete should have different periods and activities in his annual conditioning schedule. Specifically, baseball pitchers need a period of 'active rest' after their season ends and before the next preseason begins. During active rest a pitcher is encouraged to participate in physical activities that do not include a great amount of overhand throwing," the group notes.

Showcases offer young players the opportunity to display their skills to scouting services, college recruiters, and major league scouts

Chapter **10**
Pitching Strategy with Rusty Meacham

There are six ways to get a hitter out: Move the ball up, down, in, out, soft, and hard. Keep the hitter off-balance. I want to see him swinging off his front foot, because he'll have nothing on it. Go after hitters early in the count, pound it down in the zone, get ahead, and expand their strike zone.

My approach as a reliever was to watch hitters throughout the game, see where they hit the ball. I prepared myself, and when I was put in the game, I had an idea—I had a game plan on how I was going to approach them. Everybody is different. That's what worked for me.

Basically, what it boils down to is recognizing a hitter's strengths, recognizing his swings. If you sit there and you watch a hitter for a couple of at-bats, you're going to find his strengths and weaknesses. This allows you to recognize his swing when you're on the mound.

Let's say I'm facing a right-hander and I throw him two fastballs, and he hits a screamer foul down the right field line. My next pitch should be another fastball, because he's proven to me that he can not catch up to it. But you'd be surprised to learn that this basic approach still needs to be taught, even to professionals. I witnessed it all the time when I was coaching in the New York–Penn League; a guy would be in that position, he had the batter in the palm of his hand, but he'd come back with a breaking ball. If the hitter has proven he can't get around on your fastball, throwing him an off-speed pitch will speed his bat up and improve his chance of getting a hit.

So my game plan would be to study hitters.

For every series I played in the big leagues, the pitchers, catchers, and the pitching coach sat in a room and went through strengths and weaknesses of every player we would face. That's preparation—and you better believe hitters are studying your strengths and weaknesses as a pitcher.

Remember another thing: You have the baseball in your hand. I've made this mistake as a younger player: I'd throw what the catcher called for when I wasn't real sure about it, and the ball got hit hard. And I'd say, "Why'd I throw that pitch?" So I learned not to second-guess myself and think I'm always going to get beat with my best pitch. A pitcher can't doubt what he's going to throw when he throws it. Know what you're going to throw when you get that ball back from the catcher, and don't be afraid to shake him off if you don't want to throw that pitch.

Do you have a routine in the bullpen before you pitch? Or do you just get up and throw the ball to no real purpose? I'll tell you my routine if I'm a right-handed starter. I'm going to establish going away to a right-handed batter, because that's the best pitch for you to establish—away to a righty and vice versa. I go away

Every pitcher at every level of play should have a routine as he prepares or warms up in the pen. Here a Tampa Bay Rays hurler warms up in the bullpen at a spring training game. *Shutterstock*

five and then in five, away five and then in five. And then I work on my off-speed pitches.

That's a routine, a plan. I want to go out with a game plan, know what I'm going to do. I'm going to study the hitters, the scouting report. Don't be clowning around on the bench. Watch the hitters; recognize their swings and see what they do.

Throw strikes, but also have a good tempo and work fast. Hitters hate that. I would throw my pitch, get it back, and get ready to go again. Let's go. Hitters hate it when pitchers work fast.

An awful big part of pitching is in your head, not your arm. You have got to believe in yourself. You have to believe that you're better than that guy in the box. Ninety percent of this game is mental toughness and 10 percent ability. You've obviously got to have some ability, but I think the mind is the biggest thing in anything that we do. The mind controls the body.

How do you get that mental toughness? How do you prepare yourself for it? Well, we're all different. It works differently for all of us. I've just always been a confident guy. I believe in my ability. I believe that when I step on the mound I'm going to beat you. That has always been my mindset. When I was in the bullpen, I had a routine, and when I stepped out on the mound, I had a game plan. And you learned your game plan by studying hitters. I'm not one to write everything down; if you watch hitters, you're going to learn what they can and can't do.

Whatever your strength is, don't be afraid to use it against theirs. Don't think you have to do anything differently. For instance, say my strength is an inside fastball. If I'm facing Barry Bonds, even though his strength is a fastball in, I'm not going to run and hide. I'm still going to go to my strength, but I just have to make sure I put it in a good spot. You watch and you look at spray charts (charts that show where on the field hitters have historically hit the ball, usually maintained by scouts or the last starting pitcher). A spray chart can tell you whether a guy goes the other way a lot, whether he goes up the middle of the field, or whether he's a pull hitter.

That's pretty much all it takes to learn hitters' strengths and weaknesses. Then, it's up to you to attack the batter's weakness. But again, if a guy's strength is a fastball in, and I throw a good fastball in, I'm not going to go away from my strength. That's my strength and I have to live with my strength. Just like my best pitch. I'm a big fan of getting guys out with your best pitch. You can't use your second and third best.

If I'm in crucial situation in a game and I throw a slider, which is my second or third best pitch, and I get beat, I'll probably end up second-guessing myself, saying, "Ahh, I should have thrown

Four Simple Rules for Success

- Work fast
- Throw strikes
- Keep your hits below your innings
- Don't walk people

Colin Sledcik, from Haddam-Killingly High School in Connecticut, gets into the balance position. With his body in good position, Colin's eyes fixate on the target, showing tremendous focus and determination.

Three Pitches to Get an Out

Most of the parents who come to me because of my success with pitchers all want one thing from me. They all want me to increase their son's pitching velocity. It's the one thing everybody is concerned about—all the parents, all the kids. They all come to me because they want to throw hard.

Yes, velocity is very important to a pitcher. But if you don't have command and control of your pitches, you're not going to be successful. So, through good sound biomechanics, by getting your body in the right position at the right time, with rhythm, tempo, balance, and strength:

1. You'll maximize your velocity.
2. You'll increase your command and control.
3. You'll improve the movement on your pitches.
4. You'll put less stress on your arm.

That's what good sound mechanics are going to do for you. It's not just about velocity. Velocity will get you an opportunity, but it won't keep you there.

If you're a high school pitcher who throws 90 miles an hour but struggles with command and control, you *can* sign a professional contract. You're still 18 years old, so they may figure they can develop your command and control. The pros might give you two or three years to develop.

If you're a college player, and you're throwing 90 miles an hour as a right-hander, you can compete at top Division I programs. But if you're a junior or senior in college, you're not going to get drafted by the pros unless you already have command and

control, because you don't have the two or three years to develop into a better pitcher in the minors. You're 21 or 22 years old.

Let me give you a couple of examples. You're 18 years old. You get drafted in the first round of the major league draft. You're tipping 98 miles per hour on your fastball, and they pay you a $1.5 million signing bonus, but you have no command and control of your pitches. Because the investment is so large for a major league team, they are certainly going to give you the time and opportunity for four to five years to develop, because of their investment. Whereas, if you are, say, a 22-year-old senior in college and you get drafted in the thirty-fifth round, they are paying you a small investment of $1,000 a month in a one-year contract. If you are throwing 90 miles per hour, but don't have control and command of your pitches, I can assure you that you won't be coming back next year, because you are just not going to be able to perform well enough in Rookie League or Class A ball.

I always tell my guys, "If you want to be a good pitcher, you want to pitch to contact." I want to set every pitcher up with the goal of getting every hitter out in three pitches. That doesn't mean you need to strike the guy out. That means get a ground ball, get a fly ball, and the strikeouts will take care of themselves. That philosophy has been so helpful to my pitchers.

I really stress it now, because it just sends a different mentality. Instead of just throwing hard, it's pitching. It tells them they have to be in the strike zone. It tells them they don't have to be so fine around the plate. Every time a hitter puts the ball in play between the lines he is going to fail at least 7 out of 10 times. So you're better off learning to pitch to contact—unless you can throw a 97-mile-per-hour fastball, and blow it by people, and throw it for a strike. But those people are few and far between.

That goes back to the way Greg Maddux pitched, and the way Tom Glavine pitched, the way all successful pitchers throw. Three pitches to get an out. Just the other night, I looked up at a game that was in the sixth inning, with one out, and Chris Carpenter of the Cardinals had thrown 49 pitches, with two errors behind him. So really he got 18 outs with 49 pitches, that's 2.7 pitches an out. Of course, he had also thrown 15 of 20 first-pitch strikes. When you're throwing strikes, they know they have to swing.

That's the focus of every major league pitcher. That's what they're thinking: "I've got three pitches to get you out."

The Fearsome A's

One of the toughest lineups I ever faced was the Oakland A's in the 1990s. They were very, very tough. You face Rickey Henderson to lead off, then you've got guys like Carney Lansford, Jose Canseco, Mark McGwire, Dave Henderson, Mike Bordick, Terry Steinbach, and Harold Baines.

Pretty much every guy in that lineup could hurt you. You had to make good pitches, but it's very difficult to face a lineup like that. You try to challenge hitters; you work down in the zone, move the ball in and out, change speeds. I faced every one of those guys, and it was tough. I was a guy then who came in the latter part of the game, one run up or a tie game. I couldn't make a mistake. It made it even tougher for a setup man like myself to come in and face a lineup like that, where every guy can hurt you.

I think that's the difference between Double-A and Triple-A hitters, and the big leagues, where every guy can hurt you. In the minors, you might have five or six that can hurt you, whereas in the big leagues, it's the whole lineup.

I got a chance to play for Tony LaRussa, who was the manager of the A's in 1998 in spring training, a good manager and a great guy. He knows how to prepare his team. He's a Hall of Fame coach, that's for sure.

The "bash brothers," Jose Canseco (left) and Mark McGwire (above) filled in the middle of a dangerous Oakland lineup in the late 1980s and early 1990s. *Courtesy Ronald C. Modra*

Rusty pitched for the Tampa Bay Rays in 2001, his final season in the big leagues.
Getty images

a fastball in." If I get beat with my best pitch, I can tip my cap to that guy if he beats me. And he's going to, every now and then. That's going to happen. That's where the mental toughness comes in. When you play badly, when you pitch badly, you have to have that mental toughness to bounce back the next day.

Nobody Is Too Much of a Superstar

I played with Ken Griffey Jr. in Seattle. He was obviously a great player, had a great career, was a good person, and took care of the young players when I was around him. He was just a guy who was at the ballpark early every day and wanted to get better. When I played in Seattle in early 1996, he was out there taking early batting practice. He was a superstar, yet he did what he had to do. I could say the same thing for Edgar Martinez. I'd see him in spring training in Arizona at 7:30 in the morning, hitting off a tee by himself, way out there in the field, nobody else around him. Maybe that's why he was one of the best designated hitters to ever play the game.

What I am trying to say is that no matter how good an athlete you are, you need to work hard every day. You can't show up to the ballpark, put a uniform on, and expect to be good, because there will always be somebody out there working a little harder than you. So you have to outwork them. That's something Dante Bichette used to say to me. He'd look out at the pitchers and say, "Rusty, I'm not going to let that guy outwork me. I'm going to outwork him." He was a big-time player for the Colorado Rockies.

You have to play this game like it's the last day you'll ever play. When you step across those white lines, play like it's the last game you'll ever play.

Chapter **11**
The Mental Challenge

Pitching, and baseball in general, is a game of failure. You aren't going to make a perfect pitch every time, just like making an out only 70 percent of the time is considered a good thing. The beauty of the game is the challenge to overcome adversity.

A big part of baseball is controlling your emotions and staying focused on the task at hand. Routine can help. Develop routines that you can fall back on to take your mind away from any negative thoughts. Don't think, "I don't want to walk this guy." That attitude will only increase the probability of a walk, because you'll start aiming your pitches and stop throwing with the girl you brought to the party—your mechanics.

Poise is a vital quality to have as a pitcher. Everybody behind you takes notice of your visual clues. How do you handle yourself when you're on the mound? How do you handle yourself when things aren't going right? What's your demeanor? You've probably played with a guy who hung his head when somebody made an error behind him, so you've seen how bad that is for the team. You can't do that—ever.

If you make a bad pitch, ask yourself why and clear it away. There's not enough time for frustration in baseball, only time for adjustments. If you don't make an adjustment, chances are you will make the same mistake on the next pitch, and it just piles up.

This is where the ability to forge your own mental toughness comes in. Step off the mound, take a deep breath, and make an

> **The beauty of baseball is the challenge to overcome adversity**

Power of the Positive

You have to think positive. It's the most important thing we do. Think positive. If you're negative, man, you're looking at failure.

What's the toughest thing? When you fail and you're down there, near the bottom? How many people love you when you're 15–3? Everybody, right? What about when you're 3–15? Everybody hates you, right? But you have to believe in yourself. You're going to struggle. It's human. But you have to learn how to pick yourself up when you're down, and you have to learn

how to adjust. Adjustments are the biggest thing in the game.

When you pitch poorly, you have to have the mental toughness to bounce back the next day. Whether your league plays 162, 60, or 16 games, remember that it's a long season. You are not going to perform well in every single game, it's impossible. The mindset and the mental toughness to grind it out over the course of an entire season is often the difference between those who stick at the big league level and those that don't.

We've got a familiar saying: "It's tough to get there, but it's tougher to stay."

Guys are going to adjust to you. They're going to look at scouting reports, and they're going to make adjustments. You will also then have to make adjustments. Baseball is a game of mental toughness and making adjustments. If you learn how to make adjustments, and you have the right mindset, you'll last a long time in the game. Look how many mediocre players there are. They're there because they've figured it out. They've developed the mental toughness that sees them through slumps and bad days.

I wasn't a superstar. I just did what I had to do. I worked hard. I knew what I had to do. I believed in myself. During my eight years in the big leagues, I played with a lot of guys who had superior ability

but weren't able to last—they just couldn't get over that hump.

There's a quote I learned from Roger Clemens. I'll never forget it. He said: "Never be satisfied with your performance. You have to always think you could have been better." I know you've heard some things about Roger Clemens, but I'll tell you this, I don't know if I saw a fiercer competitor in my entire career. He used to pitch against us at Fenway Park when I was with Kansas City, and what a competitor he was. The other great competitor I pitched against was Nolan Ryan. You have to know how to compete. Don't think that you can beat the other guy. Know it. Be sure of it.

In the big leagues today, there are a lot of mediocre players. But they find a way to "get 'er done." David Eckstein is a great example. He's not the most talented guy in the world; everybody knows that. But he knows how to get it done. He knows what to do. What does he do when he's up there? He works that pitcher hard. He'll take a walk; he doesn't care. He makes you work. A lot of pitches are expended on Eckstein, a lot of balls fouled off, until he gets a pitcher to make a mistake, and he gets his hit. If he doesn't, he's worn the pitcher down a little.

So you have to learn to make adjustments. I am not saying it's easy. It is lonely standing out there when you're struggling.

adjustment. If you are throwing high, especially high and away, you're probably rushing your delivery, which makes you open up. Your glove-side shoulder is opening up, and you're not staying closed. Go back and throw it naturally. Stop thinking about *your* performance, and try to have fun. Remember that it's a game. You're playing this game because you enjoy it—even though you may not be enjoying the mound right now, with nowhere to hide. But getting through it, finding that inner poise, the inner strength, and going

Nick Bonofiglio, from the University of Massachusetts–Lowell, completes his throw. As he decelerates into the finish position, he must be mentally and physically prepared to react to the action of the game.

back to the way you were taught to get those last outs—that's what makes you stronger for the next time. You have to overcome adversity to know how to deal with it next time.

If you want to see how a group of men can bind together and win as a team, read Sebastian Unger's book *War*, about a unit fighting in Afghanistan. The stakes are certainly higher than in baseball, but that kind of cohesiveness really can help a team. If you're a veteran but not a leader on your team, it's time to step up if that fits your temperament. All guys are different. And as you get older, the pitchers naturally tend to congregate together and the everyday players have their club. But when you're all out there together, you're looking to pick up that other guy and play as a team to win.

Remember to have fun. You'll play better. Maybe part of the problem, is that kids don't play sandlot baseball as much as earlier generations did. Childhood is more organized, and children don't roam as freely as earlier generations. It is not this book's concern to look at changes in society, but it's hard not to notice that there is a lot more pressure put on young players who have really only known organized baseball. We try to coach players to do what comes naturally, but the lack of participation in unsupervised sports-related activity makes that more difficult.

The St. Louis Cardinals have two very good arms at the top of their rotation in Chris Carpenter (left) and Adam Wainwright (right). In 2009, Wainwright called Carpenter "the most mentally tough pitcher I've seen, been around, or even read about." *Courtesy of the St. Louis Cardinals*

Visits to the Mound

One of the funniest scenes in the funniest baseball movie, *Bull Durham*, occurs as pitching coach Larry Hockett (Robert Wuhl), comes out to visit the mound when Nuke Laloosh (Tim Robbins) is struggling. All the infield has gathered on the mound, and they each have problems that "Crash" Davis (Kevin Costner) explains, none of which have anything to do with the pitcher. Hockett solves everybody's problems, including giving a suggestion for a wedding gift. Then he exits the mound.

According to Rusty Meacham, many coaches' visits to the mound are just as inconsequential.

"Most of the time, when I'm the pitching coach, I'm going to go out there and get their mind off what they're doing, to try to clear their head. Have you noticed that most of the time, when a pitching coach goes out there, the pitcher throws a strike on his next pitch?

"I had a pitching coach who would come out and go 'Say, Jones, what did you do last night?' just to get my mind off what I

was doing and clear my head. That helped me get back to what I needed to be doing.

"Sometimes they do tell you things like 'you're dropping your arm, just get your arm up.'

"But it's almost never to get on the guy. Now, I played for a guy by the name of Lou Piniella who was sometimes hard on you when he came out there. Everybody is different, but basically coaches are just trying to calm a pitcher down, get his head clean, and get him back on track."

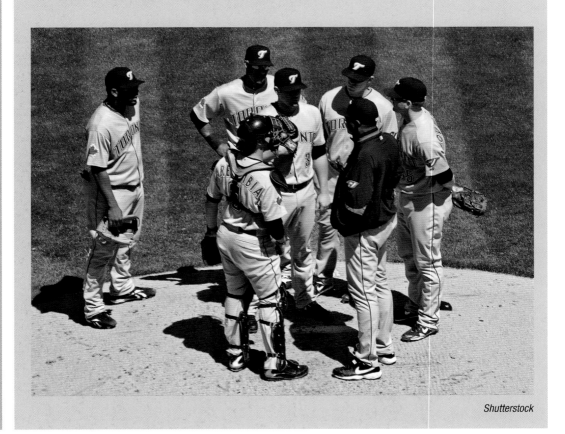

Shutterstock

Stop That Little Boy Stuff

Whoever is there watching the game—fans, family, college scouts, whoever—doesn't matter. You say to yourself before you throw that first pitch, "What do I want to be known for when this game is over?" I don't need to know what it is. You're the only one who needs to know.

Maybe you're on the mound and you realize you don't have your best stuff. Look at the opposing dugout and say to yourself, "Look, you sons-a-guns, I may not be at my best today, but I will still be here at the end of the game, and I will win the battle with whatever I have." Approach every game with that mental toughness and people will be saying "that kid's a battler out there." Your alternative is to whine and find someone else to blame when things go wrong. There's a saying I use to respond to that kind of attitude. The polite translation: "Stop that little boy stuff."

Pitching success involves repeating the same motion, and when that motion gets out of whack, it can be hard to get it back. You can't go to the bullpen and sort it out. You have to tough it out, fight your way through it. By the end of any game, you should be more mentally exhausted than you are physically exhausted, because it's simpler to get in good physical shape.

The mental part takes just as much work as your physical mechanics—maybe more. Once you get to a certain skill level, pitching becomes 80 percent mental. It's how you handle yourself, how you present yourself, how you compete on every pitch. You can't be too high, you can't be too low; you have to stay in the middle. Stay under control of your emotions, so that your body can physically compete the way it's capable of doing and so that it's nice and relaxed and flexible. If you get uptight, the body stiffens up, and you're done. You can no longer perform to your athletic potential.

You have to live in the moment of every pitch. You can't think about the pitch you just threw, or the pitch you'll throw after this one. You have to stay focused on each and every pitch. That's how you become a successful pitcher.

In any athletic activity, you will need to have medical partners who understand the nature of athletics and the ways that sports can affect the body both positively and negatively. So seek out the advice of a renowned orthopedic surgeon, experienced and accomplished physical therapists, and pitching-specific programs and plans to fortify and build your body. All the specialists in this book have treated thousands of patients and clients.

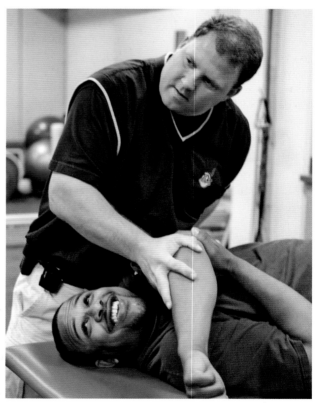

Getty Images

Chapter 12

What Does the Doctor Say?

Pain is your body's way of telling you something is broken or has broken down. It needs repair. Fortunately, the body isn't like your car; it can actually repair itself. When your car's transmission develops a burp, there isn't anything to do but take it to the shop. But a bruise on your leg will generally heal as your body repairs the damage, without your bidding. In small children, the speed of the process can be astounding. It takes about a week for a 2-year-old with a broken collar bone to heal enough to remove the yoke sling that stabilizes and holds his shoulders in place for the bones to knit. For a 16-year-old, the sling can be on for 12 weeks. The point is that when your arm hurts, rest is often all the medicine you need. Of course, there may come a time when only a doctor can fix your broken wing.

As we've said before, pitching is, in many respects, an unnatural act. Your arm wasn't designed to throw overhand. The joints in the arm are pretty amazing things when you think about it, but repeated high velocity overhand pitches can really take a toll.

According to Dr. Scott Silverberg, a prominent sports orthopedic surgeon in New York, many of his patients are in their mid-teens. The injuries that Dr. Silverberg sees most often in young pitchers are to the shoulder and the elbow.

"In the shoulder, you see tendonitis, labral pathology. You can see stress fractures or stress injuries to the growth plates.

"You see a lot of elbow pain in pitchers. That's more tendonitis or strains of the ulnar collateral ligament. When the growth plates are mature, many of them have epicondylitis. The more mature athletes start getting ligament problems," he says. "When the growth plates are open, though, that's the weakest part of the link."

> **When your arm hurts, rest is often all the medicine you need**

Growth Plates

A young person's skeleton continues to grow as he or she matures and is usually not complete until young adulthood. Growth plates are areas of developing tissue near the end of the long bones, such as the humerus, ulna, and radius bones in the arm; the scapula or shoulder blade; and the clavicle or collarbone in the shoulder. The long bones have at least one growth plate at each end. When growth is completed during adolescence—and the age of maturation can be different for any given child—solid bone replaces the growth plates.

Because growth plates are the weakest links in a young pitcher, a serious injury will more likely occur in a growth plate at the elbow or shoulder, rather than in the ligaments there. What might result in just a sprain for an adult can be a potentially serious injury for a younger person. The skeleton is still growing, much like the soft spot on a baby's head. The various growth plates ossify, or become connected bone, at different times, so there is no specific age when growth plates will disappear into bone. In general, the elbow does not mature until 16 years of age, but for some kids it comes sooner, and for others it comes later.

"It depends on the growth plate. The last ones to mature are in the clavicle—the collarbone—at about 22 years old. Each kid is a little different," Dr. Silverberg clarifies.

The Marvelous Shoulder

The shoulder is a ball-and-socket joint, similar to the hip; but, unlike the hip, the socket of the shoulder joint is very shallow and inherently unstable. It's sort of like those toys that don't quite fit together when you assemble them. Because the bones of the shoulder are not quite held in place, some sort of glue is necessary. That's where the labrum comes in.

"The labrum is cartilage, like the meniscus of the knee. It goes around the glenoid top of the shoulder," Dr. Silverberg explains.

"The labrum helps provide stability because the shoulder is built for motion. The construction is a ball with a very shallow socket. The cartilage of the labrum deepens the socket. The glenoid, the shoulder socket, is basically flat. The labrum provides a bumper."

The labrum goes around the glenoid, which forms a cup within which the end of the humerus can move. This cuff of cartilage makes the shoulder joint much more stable and allows for a very wide range of movement.

"It's got great motion. It's got more motion than any joint in the body. It's built for motion," Dr. Silverberg says.

The shoulder places the hand where it has to be

Shoulder anatomy

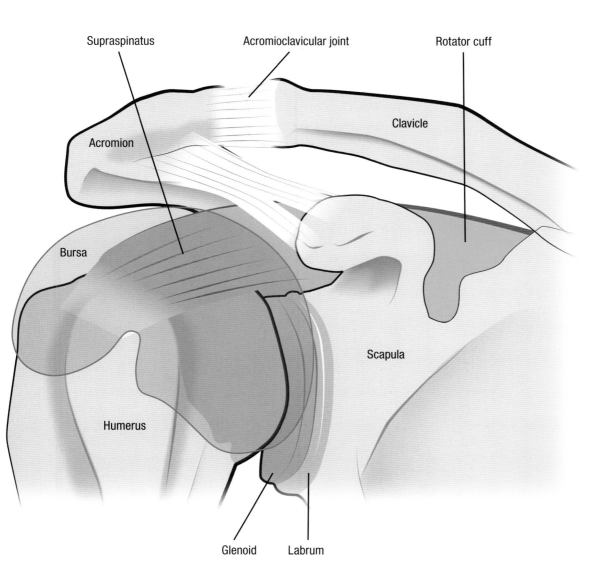

Supraspinatus

Acromioclavicular joint

Rotator cuff

Clavicle

Acromion

Bursa

Scapula

Humerus

Glenoid   Labrum

Courtesy Wes Rand

The rotator cuff, another element that enables the shoulder's movement, comprises a series of four tendons. Tendons attach muscle to bone, so the end of the muscle turns into a tendon.

"Of the four of them, the one that is stuck between the bone above and the bone below it is the supraspinatus tendon. The vast majority of rotator cuff problems are injury to the supraspinatus. That's really the only tendon in the body that is stuck between two bones. It's predisposed to having a problem. I always tell patients, 'God didn't design the shoulder so well,' " Dr. Silverberg says.

Still, the shoulder displays remarkable mechanics. "What the shoulder does, its function, is to place the hand where it has to be," the doctor explains.

Johan Santana: fierce competitor, frequent victim of injury. Santana's 2010 season was cut short due to a torn anterior capsule in his pitching shoulder. *Courtesy Ronald C. Modra*

Though we may take our ability to manipulate our arms for granted, it's eye opening to see what remarkable joints our elbows and shoulders are. Your elbow has two ligaments: one makes it go up, and one makes it go side to side. The understanding of how these parts work is called biomechanics.

"All orthopedics have to have some training in biomechanics—and they all hated it ... except for those with an engineering background to begin with," says Dr. Silverberg, who started in college at Cornell University as an engineer. "I used to teach the biomechanics portion of our orthopedic program," he says, as he chuckles over the experiences of reluctant medical students. "It was all about the fulcrums and such. When you do joint replacements, you have to understand the moment arms of the different joints."

Diagnosis

What leads an injured pitcher to an orthopedic surgeon?

"A lot of times they see the pediatrician, and the pediatrician sees they're not getting better. Sometimes, they'll see the trainer on the team, and they'll send them over," Dr. Silverberg says.

Arizona's Brandon Webb suffered a career-threatening labral tear early in the 2009 season. It remains to be seen whether Webb will be able to mount a successful comeback. *Getty Images*

A young pitcher's troubles can usually be found in the shoulder or elbow.

After learning how the injury occurred—and the kid may not know in the event of an overuse injury—the doctor performs a physical examination. The doctor will do several physical tests to check range of motion, stability, and pain, but a doctor's physical examination has limitations.

"To some extent, you can discern ligament damage from physical examination. The rotator cuff problems you get a hint at, but you don't know the extent of it until you get an MRI (magnetic resonance imaging)," Dr. Silverberg explains.

The doctor will probably use x-rays to determine if he can see any pathology and decide on a treatment plan. Because growth plates have not yet hardened into solid bone, however, they don't show on x-rays. The rim of the shoulder socket is soft tissue, so x-rays will not show damage to it. On x-rays, growth plates appear as gaps between the shaft of a long bone, called the metaphysis, and the end of the bone, called the epiphysis. If the doctor suspects something may be wrong that he can't see, he'll order other diagnostic tests, such as an MRI or CT (computed tomography). In both cases, a contrast medium may be injected to help detect a potential tear. Ultimately, however, the diagnosis may have to be made with arthroscopic surgery.

Most problems in the shoulder involve the muscles, ligaments, and tendons rather than the bones. As we've said, shoulder problems in pitchers can develop slowly through repetitive, intensive training routines, which is why a youth will seldom be able to pinpoint a specific injury occurrence. He will be more likely to describe when the pain started, and when it became uncomfortable as it gradually increased.

A lot of kids will try to play through pain: maybe because they are so determined to play; maybe because they think it's expected of them to "suck it up," in a sort of conditioned societal response to football coaching; or maybe because they don't want to disappoint their parents (and, in most cases, that means dad). It's probably some combination of all three.

The American Academy of Orthopaedic Surgeons (AAOS) cautions, "Some people will have a tendency to ignore the pain and 'play through' a shoulder injury, which only aggravates the condition, and may possibly cause more problems. People may also underestimate the extent of their injury because steady pain, weakness in the arm, or limitation of joint motion will become almost second nature to them."

So, it is important to be aware of the warning signs, and be sure to see a physician if they develop. It's your arm. If the pain

Most shoulder problems involve muscles, ligaments or tenton—not bones

is anything greater than a little soreness or fatigue, then it needs to be examined.

Symptoms of Shoulder Problems

A tear in the labrum in the shoulder causes pain, certainly, but symptoms may include other abnormalities as well.

The first indication of a tear will be pain that results from any activity where you must raise your arm above your head, such as throwing a ball or washing your hair. Pain can also come at night, especially from sleeping on the shoulder, or you may even experience pain with normal daily activities.

Your range of motion may be decreased. You may not be able to throw properly or do certain things you used to do without pain. You may hear popping or grinding sounds. The shoulder may seem to catch or lock up too. There may be a loss of strength in the arm or shoulder.

Finally, there might be a general sense of instability in the shoulder. You may feel that it just doesn't sit right, that your shoulder could pop out of its socket.

Symptoms of Labral Tear

Pain is the main symptom of a tear in the labrum in the shoulder, but symptoms can also include some other abnormalities:

- Pain that results from any activity where you must raise the arm above the head.

- Pain at night, or with simple daily activities.

- Decreased range of motion.

- Popping or grinding sounds, and catching or locking up.

- Loss of strength in the arm or shoulder.

- A sense of instability in the shoulder.

The Elbow

Let's turn to that other common sore point: the elbow. At the elbow, the upper arm bone (humerus) meets the two bones of the lower arm (ulna and radius). The elbow is a combination hinge and pivot joint. The hinge part of the joint lets the arm bend like the hinge of a door; the pivot part lets the lower arm twist and rotate. Several muscles, nerves, and tendons cross at the elbow, including the ever-pivotal ulnar collateral ligament.

The common elbow problem that doctors see in children is medial apophysitis, commonly referred to as "Little Leaguer's elbow." The more serious problem is osteochondritis dissecans.

Repetitive throwing creates an excessively strong pull on the tendons and ligaments of the elbow. The young player feels pain at the knobby bump on the inside of the elbow. This is why pitch counts are so important to watch, so familiarize yourself with the limits recommended by the American Sports Medicine Institute (see Chapter 9).

"Dr. James Andrews has done significant work there," Dr. Silverberg remarks.

"Little Leaguer's elbow" can be serious if it becomes aggravated. Repeated pulling can tear ligaments and tendons away from the bone, and the tearing may pull tiny bone fragments with it. This can disrupt normal bone growth, resulting in deformity.

"I'd say, for most kids, their problems are treatable with adequate rest and therapy. *Not* so in the elbow. They can damage the ulnar collateral ligament, which requries surgery. They can get what's called osteochondritis dissecans, which is advanced 'Little Leaguer's elbow,' " Dr. Silverberg says. "The overuse causes the death of the articular cartilage and therefore causes a loose body in the elbow. Sometimes, they need to have that piece cleaned out.

"It's probably more common in the knee. But it's always an overuse injury. I see that a lot in my kids," he says.

The AAOS says, "Osteochondritis dissecans is also caused by excessive throwing and may be the source of the pain on the outside of the elbow. Muscles work in pairs. In the elbow, if there is pulling on one side, there is pushing on the other side. As the elbow is compressed, the joint smashes immature bones together. This can loosen or fragment the bone and cartilage. The resulting condition is called osteochondritis dissecans."

"I think it's been well noted that too many pitchers throw either too many pitches in a given game or week. Either way, it's definitely too much pitching," Dr. Silverberg says.

"Little Leaguer's elbow" is the most common elbow problem doctors find in children

Tommy John surgery

Function can be restored to an elbow with an injured ulnar collateral ligament by a surgical procedure commonly called "Tommy John surgery." This procedure involves harvesting a tendon from the patient's forearm or knee and weaving it into the elbow through holes drilled in the bones.

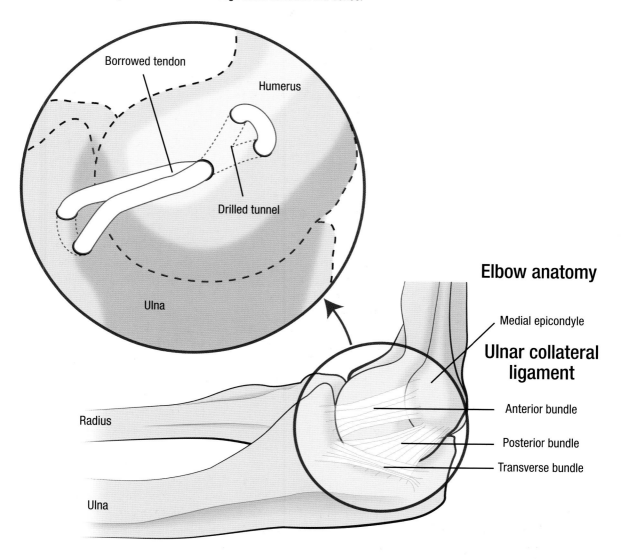

Courtesy Wes Rand

"For the elbow, I would say, importantly, that starting to throw curveballs before the growth plate is mature can lead to very serious injury," Dr. Silverberg says.

Trauma versus Overuse

You are much more likely to receive a traumatic shoulder injury in football or basketball than in baseball, where most injuries result from overuse.

"An acute labral tear causes an instability relation, where the shoulder dislocates. Even if an athlete is skeletally mature, if he has a grossly unstable shoulder, with a labral tear, you have to go and fix the labrum, otherwise he'll never throw again. But that's not a pitching injury, it's more a traumatic injury," Dr. Silverberg says.

"It's more like a fall. It's a dislocated shoulder."

So, Dr. Silverberg sees a lot more football and basketball injuries than pitching injuries: "It's just that they're more common." But traumatic injuries suffered in one sport can have consequences in another. If an athlete separates his shoulder playing football, he may become a tentative pitcher for a period of time, struggling to let loose for fear of further injury

"That's common," Dr. Silverberg says. "It's called apprehension. Once you have an unstable shoulder, you're very apprehensive about moving it in certain directions."

You might think that it's a mental problem, mostly in his head now that the shoulder has healed, but the apprehension is, in fact, a physical issue as well.

"A lot of it's physical. If you have a subtle instability, like if you have a labral tear causing looseness, then you're not going to do the things you want to do because you're feeling that the shoulder is going to come out. So, it's more than just mental. It's really a true looseness that they can't overcome."

"It's one of the most common causes for surgery in that age group."

Shoulder Injuries

According to the American Academy of Orthopaedic Surgeons, shoulder problems can be grouped into the following categories:

Instability: Sometimes, one of the shoulder joints moves or is forced out of its normal position. This instability can result in a dislocation of one of the joints in the shoulder. As mentioned above, labrum tears cause instability. Athletes suffering from an instability problem will experience pain when raising the injured arm. They also may feel as if the shoulder is slipping out of place.

Impingement: Impingement is caused by the excessive rubbing of the shoulder muscles against the top part of the shoulder blade, called the acromion. Pitching will do that, but the condition is more common in adult, professional pitchers. However, medical care should be sought immediately for inflammation in the shoulder because it could eventually lead to a more serious injury.

Rotator cuff injuries: According to the AAOS, "the rotator cuff is one of the most important components of the shoulder. It comprises a group of muscles and tendons that hold the bones

In 1974, Tommy John was the first professional athlete to undergo a successful ulnar collateral ligament reconstruction. At the time of his retirement in 1989, his 26 seasons tied a record for most in the big leagues.
Courtesy Ronald C. Modra

of the shoulder joint together. The rotator cuff muscles provide individuals with the ability to lift their arm[s] and reach overhead. When the rotator cuff is injured, people sometimes do not recover the full shoulder function needed for athletic activity."

Shoulder joint tear (glenoid labrum tear): "Advances in medical technology are enabling doctors to identify and treat injuries that went unnoticed 20 years ago. For example, physicians can now use miniaturized television cameras to see inside a joint. With this tool, they have been able to identify and treat a shoulder injury called a glenoid labrum tear," the AAOS says.

Dr. Silverberg says that a shoulder impingement injury is more likely in adult pitchers. He also asserts, "You get more rotator cuff tear in adults. A rotator cuff tear in a kid is rare.

"With shoulders, if athletes keep pitching and having pain, they tend to have more labral tears. The cartilage goes before the rotator cuff. As you get a little bit older, you can get what's called an internal impingement, or an articulus-sided rotator cuff tear. That's more in the professional pitchers, the older pitchers," he says.

"Kids tend to have more labral pathology than rotator cuff problems. If it's a rotator cuff problem at all, it's usually just tendonitis, and that's very amenable to rest and some physical therapy."

There are, of course, different levels of tears and sprains.

"With the rotator cuff, you start out with just an inflamed tendon, which is tendonitis. In the younger population, you can get tendonitis that can go on to a partial tear. Then you can go to a full-thickness tear. The same thing with the elbow. You can have a sprain or a rupture of the ulnar collateral ligament," Dr. Silverberg says.

A ruptured ulnar collateral ligament leads to what's called Tommy John surgery, named after the first major league pitcher to undergo it. The operation involves replacing the damaged ulnar ligament with one from another part of the body, typically from the forearm or hamstring. Ligaments in many places come in pairs, a redundancy that allows for continued movement when one ligament is damaged, so one can be spared for placement elsewhere. Ulnar collateral reconstruction surgery is not something that younger players are apt to get, but it is increasingly becoming a treatment for college pitchers.

Some have postulated that the surgery actually makes pitchers throw harder.

"There's no truth to that," says Dr. Silverberg. It is more likely that some pitchers throw harder simply because the regimens they follow to recover from the surgery are more rigorous than

Traumatic shoulder injuries are more common in football and basketball than in baseball

Left-hander Francisco Liriano turned plenty of heads in 2006, but his season was cut short by an elbow injury that led to "Tommy John" surgery that November. After missing the entire 2007 season and struggling for two years, Liriano was named the American League's Comeback Player of the Year in 2010. *Courtesy Ronald C. Modra*

those they followed before being injured. The pitcher is just in better shape, and he may have worked harder at his craft than he did previously.

"There are not that many surgeons who do that operation. It's a difficult operation. It takes extensive training," Dr. Silverberg says.

And although arthroscopic surgery has become more common because it is less invasive and takes less recovery time, it has a high learning curve for surgeons. The vast majority of all sports surgery now, though, is arthroscopic.

To find a doctor, consult the American Academy of Orthopaedic Surgeons, or you could ask at the hospital you go to for the name of a surgeon who treats a lot of sports-related injuries. The trainer at your school can recommend someone, or your primary care doctor or internist.

Energy Flows

Dr. Silverberg can well appreciate the kinetic flow of energy in a pitcher's motion.

"The one I remember the most is Tom Seaver, the way his body leaned in so far. So, I think, you clearly get a lot of momentum from your lower extremities. And that's why some pitchers like, oh, Oliver Perez, you don't know how much his knee problems are affecting his pitching. So there's something to be said about that," he says. "He developed a knee tendonitis, and he went straight downhill.

"You get a little less resilience as you get older. If one part of your mechanics is off, it's really going to affect the rest of you. So, that's why older pitchers are more prone to injury, because they can't compensate as much.

"They're just not as pliable. They have to stretch a lot longer beforehand to get into their routines. If they're off their routines a little bit, they're going to be more prone to having an injury."

Is there any correlation between labrum injuries and throwing across your body, rather than more vertically, and landing with a planted foot that's not straight to the plate?

"It does make sense because one of the diagnostic tests we do to see if a pitcher has a labral tear is to bring the arm across the body, you point the person's thumb down, and you have them resist elevation. If that hurts, it's an indication of a labral tear," he says. "So, a lot of it is mechanics, and a lot of people don't understand why some people get labral tears and some people don't."

Dr. Silverberg won't speculate what mechanical flaws lead to specific injuries. That's not his field. "I really treat the injuries rather than what's causing them."

Treatment

Initially, until final diagnosis, a doctor may prescribe anti-inflammatory medication and rest to relieve the pain a pitcher feels. Doctors may also recommend rehabilitation exercises to strengthen the rotator cuff muscles, if they believe that to be the issue. If these conservative measures come up short, your doctor may prescribe arthroscopic surgery.

"Surgery is necessary when you develop mechanical symptoms and are not getting better with physical therapy. So if they've torn their labrum, or they've done damage to their articular cartilage, or if they've torn part of the rotator cuff, or if they've torn the ligament to the articular cartilage," Dr. Silverberg says, pitchers will need surgery. The mechanical symptoms here are not things like opening up your shoulder; they involve the mechanical nature of the body's many parts.

According to the AAOS, "During arthroscopic surgery, the doctor will examine the rim and the biceps tendon. If the injury is confined to the rim itself, without involving the tendon, the shoulder is still stable. The surgeon will remove the torn flap and correct any other associated problems. If the tear extends into the biceps tendon, or if the tendon is detached, the shoulder is unstable. The surgeon will need to repair and reattach the tendon using absorbable tacks, wires, or sutures.

"Tears below the middle of the socket are also associated with shoulder instability. The surgeon will reattach the ligament and tighten the shoulder socket by folding over and 'pleating' the tissues."

After surgery, you will need to keep your shoulder in a sling for three to four weeks. Your physician will also prescribe gentle, passive, pain-free range-of-motion exercises. When the sling is removed, you will need to do motion and flexibility exercises and gradually start to strengthen your biceps. Athletes can usually begin doing sport-specific exercises six weeks after surgery, although it will be three to four months before the shoulder fully heals.

"Little Leaguer's Elbow"

"Little Leaguer's elbow" can lead to serious problems, including osteochondritis dissecans. And you don't want some loose bone chips floating around in a 12-year-old elbow. A pitcher needs to be shut down until he or she can recover from the pain. Continuing to throw may lead to major complications and jeopardize a child's ability to remain active in a sport that requires throwing.

> **Continuing to throw through pain may lead to major complications**

The guidelines for how many pitches a child can safely throw are listed in the chart in Chapter 9.

Symptoms

"Little Leaguer's elbow" may cause pain on the inside of the elbow. A child should stop throwing if any of the following symptoms appear:

- pain in the elbow (pretty obvious, but not always to some folks);
- normal range of motion in the joint feels restricted; or
- elbow joint locks.

Treatment Options

Left untreated, throwing injuries in the elbow can become a complicated condition. Younger children don't usually need surgery and tend to respond better to rest and a little therapy.

- Rest the elbow. Reach for things with the other arm.
- Ice packs will bring down any swelling, and for that matter, icing the elbow and shoulder after pitching if there is any swelling is always a good idea. A quick fix for minor icing: Take small paper cups, fill them with water, and put them in the freezer. After a game, just tear the cup as the ice melts and rub it along the affected areas.
- If pain persists after a few days of complete rest, or if pain recurs when throwing is resumed, stop the activity again until you get treatment from a doctor.
- Refine your throwing technique. Use the lessons in this book to improve your mechanics, and you'll have fewer injuries.

Engineering the Body

Scott Silverberg started out as a mechanical engineering student at Cornell University, but what he learned there in a couple of courses steered his post-graduate work to medicine. The body is also a machine.

"I took a couple of courses in biomechanics, and a professor of mechanical engineering, his name is Bartel, he was designing total knee replacements at the hospital for special surgery. He was instrumental in the original knee designs. I took a couple of courses with him, and I realized that all of orthopedics is engineering-based. It's stresses and strains in bones and joints. That's exactly what it is. So, this is the field of medicine that applies the principles of engineering to the body."

Little Leaguer's elbow can lead to serious injury, so it's crucial for parents and coaches to approach any of the accompanying symptoms with caution.

Chapter **13**
Physical Therapy

Rob Ackerly is that rare spectator who finds baseball a brutal, savage sport.

"When you talk about pitching in baseball, it's a pretty violent activity. For those guys to do it over and over again, it puts a lot of stress on the body," he says.

A New York physical therapist, he looks at baseball from a professional standpoint. He thinks of what throwing overhand means for the body electric, for the mechanical parts of the arm that are his area of endeavor and expertise. A therapist will stretch, mobilize, and exercise a wounded hurler's parts until he's ready to get back on the mound.

The goal is to heal and rehabilitate the injury without surgery whenever possible

Similar to doctors, physical therapists are highly trained healthcare professionals and must be certified to practice. Rob and his partner, Terry Frangopoulos, are also certified strength and conditioning specialists. Therapists who concentrate on sports-related injuries may see athletes from many different types of sports, but lessons learned about one sport will often apply to another. Swimmers can have many of the same problems as pitchers because it takes similar overhand motions to propel them through the water every day. Think about all the different sports with overhand motions; they may not require as violent a motion, but where there's a dedicated athlete, there's wear and tear. Other sports like golf, tennis, volleyball, bowling, skiing, snowboarding, and all the different extreme sports can produce major injuries to the joints and muscles. (The bowler in cricket, the man who throws the "wicked googlies," is probably also at significant risk for injury.)

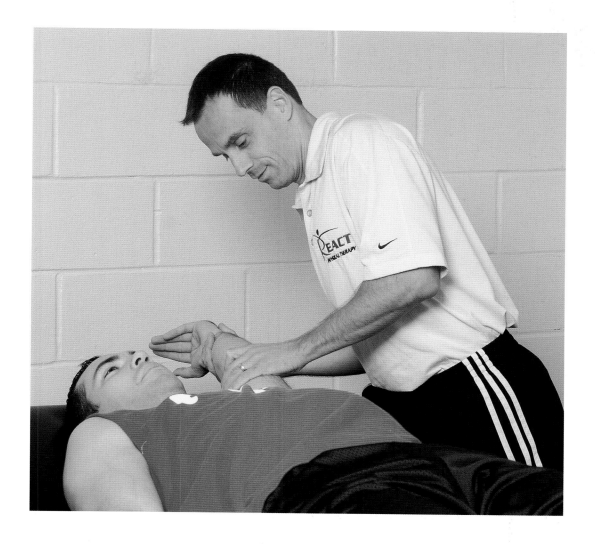

But this book is about pitching, so let's get back to baseball.

After a doctor gets a look at a pitcher with a sore arm, the next stop usually will be the physical therapist. The goal is to heal and rehabilitate the injury without surgery whenever possible. If you can avoid surgery, you want that option. But even if you are slated for surgery, you may see a physical therapist for an evaluation, although the doctor has already done some of the tests that a therapist will do.

Rob says, "When somebody comes in, we'll evaluate them. If something is tight, there are some techniques we can use to actually start some range of motion. We'll get into some more stretching if they need it.

"Range-of-motion exercises just involve the joint itself. How far does the joint move, whether actively or passively? Stretching,

Physical therapist Rob Ackerly works through range-of-motion exercises with a patient.

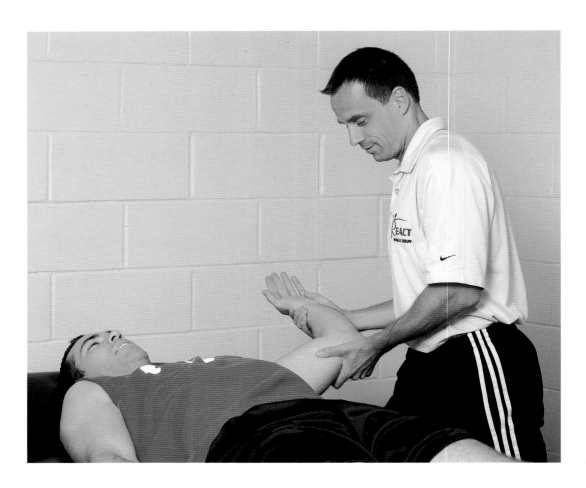

Using a particular pattern of motion, the therapist can tell how strong the patients are and whether or not there is any pain through the range. From there, a therapist puts the patients on a program where they can do the exercises themselves, whether it's with resistance bands, with weights, or in a gym.

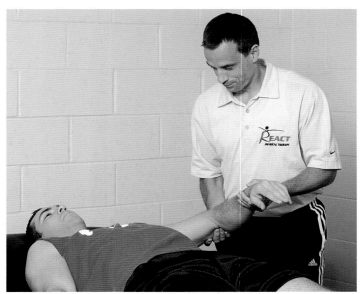

then, starts to involve structures that you're trying to identify that are tight, ones that you can start to stretch."

Like any good mechanic, a lot of what a physical therapist does is to loosen overly-tight bolts—joints and muscles in the human body. A key word for a physical therapist is *tightness*. It's what they look for; it's what they fix.

The majority of injuries that Rob sees in pitchers are in the shoulder and the elbow. Injuries to young pitchers typically are tendonitis and tears in the labrum, and in older pitchers, maybe some rotator cuff issues.

Tightness in the Joint

After evaluating a patient, a physical therapist designs a treatment program. Different injuries call for different combinations of therapy. Key components in any rehabilitation plan include education about the cause of the problem and exercise to improve range and function. Exercises are designed to improve range of motion, muscle strength, balance, endurance, and coordination. In addition to exercise, a patient may be treated with massage therapy, traction, heat therapy, water therapy, and other specialized techniques.

Manual work that a therapist can perform often involves stretching: the muscles, the joint itself, or capsules that surround joint components. The capsule of the joint can get tight, and muscles can get tight. The therapist will work to identify areas of tightness and begin to stretch whatever needs to be more flexible.

"There are many ways to stretch the parts of the body to improve it," says Theresa Solesky, Len's sister and a physical therapist in Fort Pierce, Florida.

"We'll look into any joints that seem tight," Rob says. "You evaluate it by trying to move it. If it doesn't move, then we'll try to mobilize it."

Therapists use various grades of mobilization. After that, a therapist does some manual stretches and gives patients some stretches they can do by themselves. "There are some techniques, though, that we do, that they would have trouble doing themselves," Rob says.

Joint mobilization happens, for the most part, before surgery, but it can also be done post-surgery after a certain time frame.

"After about six weeks, you'd start doing some mobilizations on these guys. A lot of guys come back after surgery and make the range we're looking for pretty quickly," Rob says. "Or, you might check their range of motion before surgery and find it's pretty good. If it's not, we'll be doing some manual stretching, some range of motion techniques."

The Body Is a Chain

Sometimes, people come straight to the physical therapist without visiting a doctor. Sports team trainers may send them, or the client may simply be looking for some answers.

"We get referrals by word of mouth, and we have a gym here, but a majority of our patients come from doctors," Rob says. "We'll see someone in the gym who may have hurt himself. If someone sprains an ankle he may come in to us, and we'll refer him to a physician to get a prescription for therapy. It's tough nowadays, but we like to coordinate everything with the doctor. Make sure he knows what we're doing."

In any case, the patient receives a thorough examination.

"If somebody comes in with a sore shoulder, we put him through a battery of tests," Rob says. "There are special tests for the shoulder, special tests for the elbow, to pinpoint certain structures. You can do certain rotator cuff tests; you can do certain impingement tests. It can help us understand whether it's a rotator cuff tear or just a rotator cuff impingement, versus a labral tear—before just telling them 'go get an MRI.' "

Rhythmic stabilization exercises for the shoulder are typically used for just about anybody that comes in with an upper quarter injury. The therapist applies gentle resistance in opposite directions. This allows for gentle activation of rotator cuff muscles.

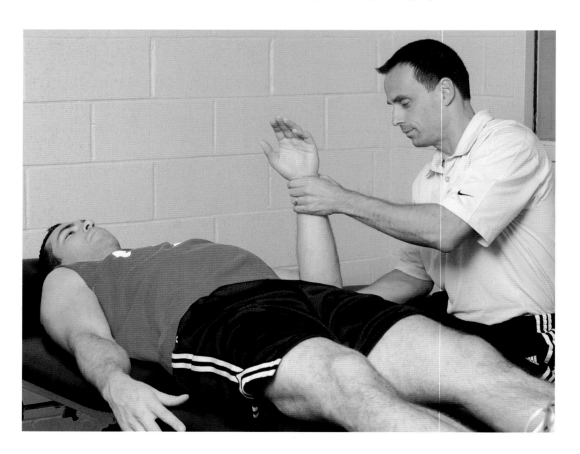

The next move is to go through strength testing. The therapist assesses the different muscle groups that are affected by the injury or may even have contributed to it.

"If we're looking at a shoulder injury," Rob says, "we'll treat the elbow too. If it's an elbow problem, we're not just going to look at the elbow; we're going to look at the whole chain.

"The body's a chain, and we're going to see if there are any weak links that we can find. Normally, we'll look at the shoulder for wrist and elbow injuries, and we'll do some strengthening or range of motion, whatever needs to be done. We're looking at that in any upper quarter injury that we see," Rob says.

An experienced therapist can usually get an educated idea of the severity of the injury—whether or not the client is going to need surgery. Ultimately, magnetic resonance imaging (the MRI) is going to be the best way to determine severity conclusively.

"Not that an MRI is going to be 100 percent on the money," Rob says. A surgeon can begin to operate and find the injury is not that bad, or maybe it's even worse or more complicated than the MRI indicated.

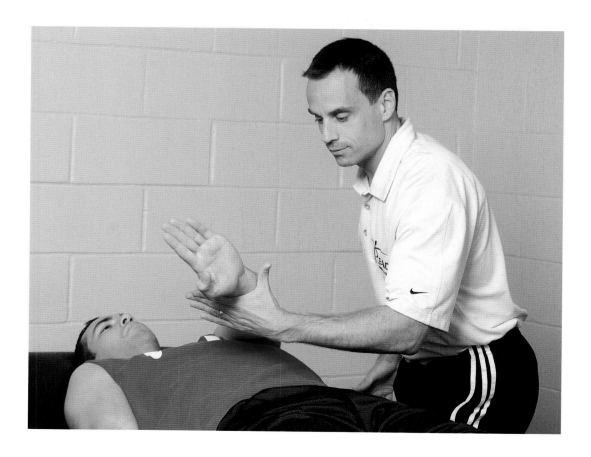

Rehabilitative exercises are designed to improve range of motion, muscle strength, balance, endurance, and coordination.

"The MRI is always going to be something that these guys get, to get a closer look, take a peek under the hood, so to speak, and see whether it looks like the patient has a complete tear, or how large is it," Rob says. "For us, we can tell. If it were something as severe as a complete tear, we'd refer him back to the doctor. If they've come from the doctor, then the doctor is going to have a pretty good idea. He's done his battery of tests, and he'll have a good idea of what he thinks it is.

"We'll go through our battery of tests, so we're both on the same page, what we find, how we're going to treat it. You get an idea of whether the patient is going to improve without getting surgery or if surgery is the best option."

Strength Testing

In the strength-testing process, techniques include manipulation of the joints and resistance exercises.

"We'll assess, and it'll tell us," Rob says. "A lot of times, if it's weak and painful, then there's something going on with the muscle itself, maybe a rotator cuff tear, or at least a strain there. If it's just weak and painless, then it's just a matter of weakness, not a tear, because there is no pain involved. Based on that, we'll treat accordingly.

"It's always a constant evaluation, as you progress these guys in their rehabilitation, in an effort to ultimately figure out what is going to help these guys out."

Previous shoulder separations can hinder an athlete. "That shoulder, the AC joint, is key. Major instability there can affect the entire chain. That's certainly going to magnify issues mechanically in the shoulder and the elbow," Rob says.

But sometimes, pitching the right way can still lead to trouble.

"If this guy, say, has perfect mechanics and can heal quickly and is tough, on the tougher side, and doesn't break down easily, you still can have some overuse issues, regardless of whether they start out great—or if they have an injury in football and it carries over to baseball," he says.

Instability in the AC joint can affect the entire chain, magnifying issues in the shoulder and elbow

Pitchers' Needs

Physical therapists can even improve nerve damage. For instance, the ulnar nerve in the elbow can become a source of pain, though principally for older pitchers.

"We can do certain things to position to see if they have some tightness along that chain, the one the ulnar nerve comes from. There are actually some neuro tissue mobilization techniques we can use, if that's necessary," Rob says.

"Once you have an injury, you're going to have some inflammation. It's going to start scarring, create scar tissue, the process of scarring itself. The thing is going to scar no matter what you do. Our job becomes . . . to minimize the scar tissue, or at least make it functional, make it not impede function," Rob explains.

"If things scar down, you can have adhesions, things become tighter. You're definitely going to want to mobilize that neural tissue. There's a sleeve basically around the nerve, and that can get a little tight. Just like you can stretch a muscle, you can stretch that nerve tissue a little bit. Usually, with the ulnar nerve, it's ice, rest, and then slowly progress him back, get some full-range motion back—and then start some strengthening."

It's all part of the whole, just like the correct way to pitch a baseball.

"The big thing for us is if we know what positions athletes are supposed to be in, we can look and see what's tight," Rob says. "Even with baseball pitchers, you see their legs, there's some pretty good leg movement too. So we look at their legs. We'll look at the hip. We'll look at the ankle. We'll look at the knee. Are there any issues where your hip flexor is tight? It can affect the chain. It's putting more stress on the upper body, putting more stress on the shoulder, putting more stress on the elbow. "So, it certainly can be a factor."

The need for an overall view is just more pronounced in athletes who participate in activities that require extreme motions.

"So, we'll look at all those joints, all of them, to see what might be tight, what might be holding them up," Rob says. "We want to make their mechanics better by improving their flexibility."

"Those are your two biggest things: improving flexibility and strength," says the physical therapist.

"The first thing is: We're going to get them as strong as possible, see if there are any glaring weaknesses anywhere, or things that need to be corrected to make their mechanics much better. We're going to find things that need to be corrected," Rob says.

If you go from joint to joint, it's the same idea.

"If we looked at somebody, and let's say he was going back into the cocking phase of the throw. Maybe when we evaluate him, we see that there's some tightness somewhere, whether it's petrallus major or it's a Little League shoulder joint issue, " Rob says. "When it's tight, something else has to move quicker, sooner than you really want it to. That can certainly affect the medial elbow joint. So here's a good reason to look at the shoulder because maybe this is the reason his elbow is not doing great."

Chapter 14

Twenty-first Century Strength and Conditioning for Pitchers

By Eric Cressey, MA, CSCS

In the past, strength and conditioning for pitchers was an afterthought—if that. Sadly, there was—and continues to be—a school of thought that insists that there is no place for strength and conditioning in this population because "they're not athletes; they're pitchers." Where has this mindset gotten the baseball industry?

Professional teams waste *hundreds of millions* of dollars each year on injured pitchers on the disabled list. And orthopedic surgeons and rehabilitation specialists make *hundreds of millions* of dollars on elbow and shoulder injuries in youth pitchers who are flat-out unprepared.

The research is pretty clear that—in addition to flat-out overuse—being physically unprepared is a primary reason pitchers get hurt. And it shouldn't come as any surprise, given that we know throwing a baseball is the single fastest motion in all of sports. Sending an out-of-shape pitcher out to try to throw 90 miles per hour is like asking your beat-up old car with 300,000 miles on it to go out and compete in the Daytona 500; it just won't end well. An effective pitcher needs a unique blend of strength, power, mobility, and soft tissue quality to perform at a high level and stay healthy in the process.

Fortunately, baseball is slowly emerging from the dark ages with respect to physical preparation of players. Unfortunately,

> **Baseball is slowly emerging from the dark ages with respect to physical preparation of players**

old mindsets don't disappear overnight, so we're left with a lot of poorly thought-out traditions that impair the developments of a group of athletes who desperately need effective programming.

While I certainly won't be able to teach you everything you need to know about strength and conditioning for pitchers in a single chapter, I will highlight five key messages that I think every pitcher needs to understand. Armed with this knowledge, you'll be able to get started on the right path and be an advocate for your own development moving forward.

Message No. 1: Strength Training Is Not Optional

As this book has gone to great lengths to show you, pitching is all about producing skilled movements in the right sequence. If your mechanics are flawed, all the strength and flexibility in the world won't do you any good.

Interestingly, though, what many players and coaches fail to realize is that having inadequate strength/stability (and mobility) can actually make it impossible for a pitcher to get into certain positions to allow for the mechanics that are right for him. Moreover, being weak and inflexible are recipes for disaster on the injury front—and an injured pitcher will never be able to repeat his mechanics because he won't be throwing! Resistance training is the basis for modern physical therapy, which is aimed at restoring appropriate movement patterns and alleviating symptoms. If it's appropriate for injured individuals, it has to be beneficial for healthy athletes.

In making the case for strength and conditioning, I'll be very blunt. On *countless* occasions, I've seen pitchers gain velocity without making any changes to their throwing programs or mechanics. I know what many of the devil's advocates in the crowd are thinking: "You're just making that up!" So if my word isn't enough, how about we just go to the research?

Dr. Copp Derenne and others make it quite easy for us, as they published a great review of the effects of resistance training on baseball throwing velocity in the *Journal of Strength and Conditioning Research* in 2001. The authors reviewed 26 studies that examined the effect of different strength protocols on throwing velocity, and 22 of the 26 showed a favorable effect in those who resistance trained; the improvements were significantly greater than those seen in controls who just threw. In other words, throwing and strength training are better than throwing alone for improving velocity—*independent of optimization of mechanics from outside coaching.*

Yes, this review was published in 2001—meaning that some of the "old guard" has continued to fight the utilization of strength training in pitchers for a full decade in spite of the fact that there is a ton of scientific evidence to justify its use. If you want to weed out a bad pitching coach, ask him what he thinks of strength training. If he shoots it down, find yourself a new coach.

Taking things a bit further, in my opinion, the training programs referenced in this review were actually not very good protocols. Most relied heavily on isolation exercises and higher rep schemes. Some only used tubing and light dumbbells—far from a comprehensive

Lunges are a staple exercise, and adding dumbbells can improve results.
Shutterstock

strength training program. If archaic stuff works, just imagine what happens when pitchers actually train the right way—and have pitching coaches to help them out to use this new strength and power on the mound!

So what does a "good" strength training program for pitchers consist of? It's a difficult question, for sure, and you might get 10 different answers if you ask 10 different strength and conditioning specialists. For that reason, I encourage you to seek out someone qualified in your area to supervise you in pulling together a comprehensive program and to coach you through proper techniques.

With that said, here are some guidelines for getting things in order:

1. **Strength and conditioning are about more than just lifting.**

   We aren't just talking about lifting weights. A comprehensive program should also include a thorough warm-up and a dedicated approach to flexibility, medicine ball, and speed training.

2. **Focus on compound, multijoint exercises.**

   For example, you're better off doing a dead-lift variation than you are a leg curl because the dead-lift involves multiple joints and builds much more muscle mass. A chin-up would be better than a biceps curl for the same reason. Sure, there is a place for isolation exercises like rotator cuff drills, but the bulk of your training should focus on the exercises that will give you the most "bang for your buck."

The dumbell bench press is preferred to the straight bar variation.
Shutterstock

3. Omit certain exercises.

While there are quite a few similarities between how I train my pitchers and how I train my other athletes, there are several exercises we omit with our overhead throwing athletes. Contraindicated exercises in our baseball programs include:

- Overhead lifting (chin-ups are okay)
- Straight-bar bench pressing (we use dumbbells)
- Upright rows
- Olympic lifts
- Back squats (we front squat instead)

4. Focus on the "staple" exercises.

Here's a small list of some of the exercise variations used most frequently for pitchers in our strength training programs:

- Front squats
- Dead-lift variations
- Single-leg exercises (lunges, split squats, single-leg dead lifts)
- Sled pushes/pulls
- Slideboard drills
- Push-up variations
- Dumbbell bench pressing variations
- Every row and chin-up variation you can imagine (excluding upright rows)
- Thick handle/grip training
- Medicine ball throws
- Conventional rotator cuff exercises with bands and dumbbells
- Various core exercises, beginning with bridging variations (not sit-ups and crunches!)

Of course, not every one of these exercises is appropriate for each pitcher—and, of course, you don't want to do every one of them in each session!

5. Never train through pain.

This seems like an obvious statement, but it warrants mention just because so many pitchers foolishly throw and train through pain because they think it's normal to hurt. Pain is a sign that something is wrong, and you need to rearrange your training to avoid the issue. See a qualified professional if you are having pain with throwing and/or lifting.

There are countless variations you can use to make your push-ups more effective. *Shutterstock*

Message No. 2:
It's About More Than Just the Throwing Program

During the season, the throwing program could be considered the center of a pitcher's universe, as it dictates the schedule—from starts, to bullpens, to long toss. It tells a pitcher where he needs to be on a given day and what he needs to do. It seems pretty straightforward, right? Wrong—because there's more to complete pitching development than just throwing!

The Five-Day Rotation

In a case of a five-day rotation for a starter (usually reserved for professionals), it's difficult to get in a ton of quality non-throwing work between starts. Keep in mind that dynamic flexibility and static stretching are performed every day, and core work is included in some capacity in every strength training session.

Day 0: Pitch

Day 1: Lower body lift, light push-up and rowing variations, light rotator cuff work

Day 2: Movement training only, focused on 10–15-yard starts, agility work, and top-speed work (50–60 yards)

Day 3: Single-leg exercise, upper body lift, rotator cuff work

Day 4: Rest (or just some low-intensity dynamic flexibility circuits)

Day 5: Next pitching outing

The Seven-Day Rotation

With a seven-day rotation, we've got a lot more wiggle room to get aggressive with things. This is why in-season can still be a time of tremendous improvements in the college game, especially since you can work in a good two or three throwing sessions between starts. Again, dynamic flexibility and static stretching are performed every day, and core work is included in each strength training session.

Day 0: Pitch

Day 1: Lower body lift, light rotator cuff work

Day 2: Movement training only, focused on 10–15-yard starts, agility work, and some top-speed work (50–60 yards); upper body lift

Day 3: Extended dynamic flexibility circuits

Day 4: Full-body lift

Day 5: Movement training only, focused on 10–15-yard starts, agility work, and top-speed work (50–60 yards)

Day 6: Rest (or just some low-intensity dynamic flexibility circuits)

Day 7: Pitch again

These schedules aren't perfect for everyone, so you'll probably find that you need to tinker with them as you find what works best for you. And, if you're a relief pitcher, it can be difficult to get things in, so I encourage you to get in your quality lifting right after a shorter appearance in case you need to be ready to throw again shortly thereafter.

Shutterstock

The Off-Season

As a strength and conditioning coach, I love the off-season because it's when the throwing program can take a backseat for a few months while we build up the athlete. Here's what a typical week might look like for my professional pitchers:

Monday: Medicine Ball / Mobility, Lower Body, Rotary Stability
Tuesday: Upper Body, Anterior Core
Wednesday: Movement Training, Medicine Ball / Mobility
Thursday: Lower Body, Rotary Stability
Friday: Medicine Ball / Mobility
Saturday: Movement Training, Upper Body, Anterior Core
Sunday: Off
Daily: Individualized Flexibility and Soft Tissue Treatments (foam rolling, massage, etc.)

As the off-season throwing program begins, we lower the training volume so that "specificity" can take over again.

Young athletes, of course, have a different set of demands— from schoolwork to other sports. You don't have to have a crazy training schedule like this to make progress, but you do need to be consistent throughout the entire year to experience the great benefits strength and conditioning can offer. For most young athletes, three full-body training sessions per week—plus some additional movement training, whether it's doing agility and sprinting or just playing basketball—will get the job done during the off-season. During the season (whether it's baseball or another sport), just getting in two shorter resistance training sessions per week will go a long way.

Message No. 3:
Distance Running Deserves a Place—
in the Garbage

I was put on this Earth to put an end to distance running for pitchers. It just has never made sense to me. While I could hop onto my soapbox and spend 15 pages outlining everything that's wrong with using distance running in a pitching population, I've already done that—so there is no sense in reinventing the wheel. Head over to www.ericcressey.com and search for my two-part series of articles titled "A New Model for Training Between Starts." I think you'll find both to be very entertaining, and I'm positive that you'll see tremendous improvements in pitching performance by incorporating the suggestions I make in those articles in lieu of distance running.

Message No. 4:
Flexibility Is Everything

It goes without saying that flexibility is huge for pitching performance. However, what a lot of people don't realize is that it's also imperative to maintain flexibility in order to maintain arm health. The two biggest flexibility deficits we see in pitchers that subject them to arm issues are a lack of shoulder internal rotation and a lack of elbow extension. Pitchers can quickly lose these ranges of motion because of the tremendous stress that arises during the deceleration phase of throwing. Here are three stretches that can be used to address these two issues:

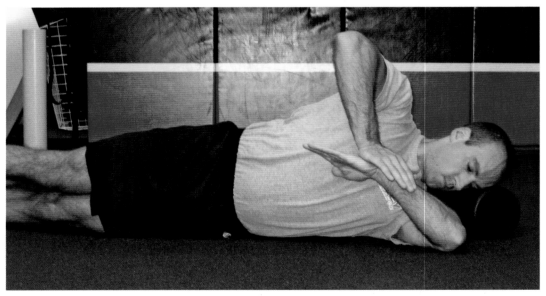

Sleeper stretch (throwing shoulder only). Be sure to keep the shoulder blade locked down and back to stabilize it against the ground as you perform this movement. If you can touch your hand to the ground, you are not doing it correctly. Head support is very important, so if you're at the field, just put a ball in your glove and use that. This should be a very gentle stretch; don't crank on your arm like crazy!

Elbow/wrist stretch 1. The goal here is to straighten your elbow while extending your wrist and supinating your forearm.

Elbow/wrist stretch 2. This stretch is the same as the previous one, but with wrist flexion and forearm pronation.

Wall lat stretch. Many pitchers are limited in how well they can flex the shoulder. In other words, they can't get their arm over their head very well—not exactly ideal for a pitcher! Try this gentle stretch for the area, where you use your opposite hand to pull your shoulder blade down toward your hip.

Obviously, lower body flexibility is of paramount importance, too. Try these stretches to work on that.

Split-stance kneeling adductor stretch. You need adductor (groin) mobility to get down the mound, get the most out of your push-off, and shorten the distance between you and the batter. Perform this stretch in both hip extension and hip flexion on both sides, as pictured at right.

Lying knee-to-knee stretch. You need hip internal rotation to be able to "swivel" over your front leg during the follow-through. You should feel a stretch in the outside of your thighs and glutes.

Kneeling heel-to-butt stretch. This stretch gets both the quadriceps and hip flexors, allowing you to get down the mound by fully extending your hip.

Three-way band hamstrings stretch.
This three-position stretch for the hamstrings will allow you to really come over the front leg and get on top of the ball. Limited hamstring flexibility is a common culprit when you see a pitcher leaving pitches up in the zone.

The Take-home Messages

As I noted, this was just a brief glance at some of the key factors that make or break a pitcher's strength and conditioning program. At the very least, I hope you have grasped that there is nothing fundamentally wrong with strength training programs for pitchers. In reality, what is wrong is the assumption that all strength training programs are useless because some are poorly designed and not suited to athletes' needs and limitations. Be leery of people who say strength training isn't important. Everyone—from endurance athletes, to grandmothers, to pitchers—needs it! And its results are magnified when combined with the right flexibility, throwing, and movement training programs.

For hundreds more detailed articles on all different facets of training for improved athleticism and injury prevention on the mound, please visit the "Baseball Content" section of my website.

About the Author

Eric Cressey, MA, CSCS, is president and co-founder of Cressey Performance (CresseyPerformance.com), a facility just west of Boston, Massachusetts. Eric has helped athletes at all levels—from youth sports to the professional and Olympic ranks—in a variety of sports, and he is perhaps best known for his extensive work with baseball players. Cressey Performance has rapidly established itself as a go-to high-performance facility among Boston athletes, as well as for those who come from across the country and from abroad to experience CP's cutting-edge methods. The author of five books and more than 300 articles and creator of four DVD sets, Eric publishes a free newsletter and daily blog through his website, EricCressey.com.

PART 4
WHAT NEXT?

You've made it through high school ball and you're a star (or not quite) on your Legion team as a junior member. There's somewhere you can play if you are determined to continue to play baseball in college. This section shows you ways to find the school and baseball program for you.

Shutterstock

Chapter **15**
Choosing Your Post-High School Path

The Dilemma of Sport Versus Education

There is always a college where you can continue to play baseball—if you are talented enough and if you truly want to play baseball. Florida itself is full of them. But you have to be committed to the game like you have never been before. It's an adult pursuit, because you are now in an adult world. But when picking a college or university, you have to weigh whether the scholastics at the school best fit your later needs in life, because the likelihood that you can make a living playing baseball is not great.

We recommend that college-bound students choose five schools—two dream schools and three realistic schools. By the time you are a junior in high school, you should know what your capabilities are and this should help you determine the division level of colleges you target. Most high school kids have a pretty good idea where they fit, though some kids may be "tweeners," kids who could play in either of two divisions.

The most important thing, though, when a young pitcher is picking a school is the prospect of his academic career. The reason is obvious: Only 6 percent of first-round draft picks ever step on a major league field. So academics should certainly be a priority when picking a college or university.

You want to pick out a school that fits you academically and in terms of baseball. If you're a pitcher, and you're satisfied with a school's academics, you want to look at the baseball program: the coach's history, the overall coaching. What kind of reputation does the pitching coach have? Are you going to benefit from the pitching coach there and the baseball program? Now, if that works, then you go to Plan B, and that's your second college choice. Do this for your five choices, though you may have to investigate more than five schools to pick five you're comfortable with.

That's how you want to look at it: Academics first, baseball program second, and then, of course, the fit of the location, where you want to be. Close to home, or maybe farther away? Do you like the big city or are you more comfortable in a quieter location? You have to take that into account as well.

In Division I, there are higher-level baseball programs—schools such as Florida State, the University of Miami, UCLA, Oregon State, and Wichita State. These teams are in the Southeastern Conference, the Atlantic Coast Conference, or the Pac-10, where the competition is fierce. Because they are high-level Division I programs, there are certain criteria their coaches are looking for. Probably the number one priority for them is velocity. Division I programs like Florida State and Miami want right-handed pitchers who throw at least in the high 80s.

Then there is a middle tier Division I schools—places like Northeastern University or University of Hartford, Dartmouth, Harvard, the Ivy League schools—great schools, but not top-25 baseball programs. If you fit into that category, you might only have to be in the 83- to 86-mile-per-hour range. But you better have command and control of your pitches. There are also some lower Division I programs, schools such as St. Mary's and Sacred Heart.

In Division II, there is an even greater variety of leagues. In the Northeast, you have a Division II league that's a wooden bat league. You don't need to throw 88 to 90 miles an hour in that league; you can pitch at 81 to 84—if you have command and control—and be very successful. Or you can go into a small Ivy League–like school with aluminum bats, such as Amherst, Trinity, or Williams. They are pretty good Division III programs, but again, you better have command and control of your pitches. That's what it takes to pitch in Division III; you don't have to throw 85 or 86 miles an hour to be successful, as long as you can throw strikes and get batters out.

After that you have your junior colleges—for instance, the many Florida junior colleges, which basically aren't the best academic institutions. But there are a lot of kids out there who know that what they want to do is play baseball for a living. So if you want to play baseball for a living, in Florida, you can go play the equivalent of a high-level Division I baseball at a junior college level. It's where a lot of high-level D-I schools look for players to transfer into their programs. A majority of the junior college players in the southern part of the state get drafted by major league teams. That's pretty much what they're down there for, to play baseball, and academics aren't that important to them. So that's not for everybody, but for some, it's what they want.

In college, always keep your priorities straight: academics first, baseball second

Pro Contract Versus College

Want a big decision to make? After high school you are drafted by the major leagues—but you also have college scholarship offers. Your decision depends on your goals and dreams.

So you're a junior in high school. You have just finished all of your summer ball, tournaments, showcases, etc., and Vanderbilt University just offered you a scholarship for baseball. You've had thoughts of being a doctor, of enrolling in Vanderbilt's pre-med program, and you committed to the scholarship. A year goes by. It's the June draft, and you're drafted by a major league team in the second round. You get offered $750,000 to sign, and you would also save the cost of four years of college if you opt to accept the offer.

This is a big decision. Of course, it all depends on your circumstances.

If you choose to go to college, the worst thing that could happen is that 1) you hurt your arm severely and are never going to develop the potential you had (you lose that dream to ever step on a major league field); or 2) you don't develop your potential in college and your stock in the draft will drop considerably.

You might have to settle for a lot less money. Or you might not even get drafted. The opportunity has flown away.

On the other hand, if you graduate from Vanderbilt University you can't put a price on that education. You will carry that with you for the rest of your life.

Now, let's consider that you made the decision out of high school to sign a pro contract. They assign you to, say, the Gulf Coast League in Florida, Rookie League ball, short season Class A. You find yourself pretty much in a 9-to-5 job situation. You're working now. You're an adult in an adult world. But you're only 18. You find out that everybody has stuff as good as yours. You feel like a pebble in the sea, and you're really starting to understand that it is going to take a whole lot of hard work to ever step on a major league field. Are you really prepared for that? Can you handle that? Do you have what it takes to step on a major league field? You'll be thinking about that; trust us.

Remember, only 6 percent of number one draft picks ever get to the show. Do you have what it takes?

These are just some things to think about, if you ever get lucky enough to have that dilemma.

Perils of Athletic Department Politics

What happens if the college coach who recruited you and convinces you to attend XYZ U is fired or jumps ship for a better job? Be aware of some of the hazards of collegiate athletics. These guys are all adults now and the jobs aren't always the fulfillment of their career ambitions or their athletic director's. They need wins to keep their jobs. Baseball isn't a dream for them, though they make their living at it. When you go to college, you are living in an adult world, with all its compromises and broken promises.

Some college coaches may "guarantee" playing time to promising players they want to recruit. They know as well as anybody that nobody wants to ride the bench, even if he does have a scholarship (which, in some situations, he may lose after one year if he's not playing). Be wary of those kind of promises, because you will always have to earn playing time.

Sometimes it seems like a coach favors you before you've even committed to play for him. Of course, that's not always the case or the rule. For instance, a lot of coaches will tell you straight up whether they need left-handed pitching or not.

Coaches also get fired, or "retire." So the guy who recruited you and put you on his team may not be the coach next year, and the new guy may not be as impressed with your skills. He may have other priorities, a different philosophy or approach, or he may just fall in love with another guy. (Though they're better, college coaches can suffer from the same things all athletic coaches do—which is not seeing things the same way you do.)

Programs have also fallen victim to budget cuts. In recent years, the University of Vermont had a very good baseball program and good coaches, only to see it lose out to other university priorities. The University of New Hampshire and Providence College, schools at which baseball was the first varsity sport, both cut their baseball programs in recent years. Baseball does not produce revenue for the school like basketball and football do, so you should remember it can be a vulnerable program.

Getting Scouted

How do you get yourself noticed? One of the most successful ways today is to play with an AAU team. These teams compete in national tournaments that draw college and professional scouts and scouting organizations. It works somewhat like the NFL scouting combines—the scouts don't have to go all over the country to see the best players.

Perfect Game, the largest high school scouting service in the country that colleges use to recruit players, holds these national tournaments. They rate players in the tournaments, so the colleges have all the kids' statistics and background online to research the players. At the tournaments, or showcases, college recruiters get to see 220 to 230 teams at one time, at one location, with generally the best players in the country. So it's advantageous for you to get on an AAU program team that attends those tournaments, so you get exposed to the recruiters who attend the showcases. These teams are also very competitive, so you want to play at your highest possible level.

American Legion baseball is very competitive, and a lot of local colleges will attend those games, but you don't get the national showcase exposure. It's easier for recruiters and scouts to see everybody in one location, rather than spread out all over the country. They just don't have the time, because they're playing 52 games a year and practicing year-round. If they didn't have these showcases, the only way you could be seen is pretty much by word of mouth. So we highly recommend that kids play in a high-quality AAU program that attends national Perfect Game showcases, which is the largest recruiting tool that college coaches use today.

Playing on an AAU team is one of the best ways to be seen by scouts

There also are a number of other things you can do to market yourself.

- One of the most important recruiting tools today is the skills video. The video should be edited, last about two minutes, and be posted on the Internet. Find out the baseball coach's e-mail address from the website of the university you are interested in, and send the coach an e-mail introducing yourself, with a link to your video. YouTube offers a pretty easy path for putting up your video. A video is a critical element in college recruiting because it allows coaches to get a sense of a player's talent with very little effort on the coach's part. Be sure to follow up, but give the busy coach time to look at what you have. Do this in the college off-season; during the

Since his retirement as a player following the 1984 season, former New York Mets catcher John Stearns has been active as a scout, instructor, coach, and manager.
Shutterstock

On the Big Leagues

On a typical game day, I would have to there about one o'clock in the afternoon for a 7:05 game—this is as a coach. The time depends on whether you are on the road or at home, because at home you take batting practice first.

With a seven o' clock game time, as a player, you're there dressed and ready at four o' clock. You go out and do your stretching. Do your long toss if you're a pitcher; same thing if you're a position player, do your playing catch. Then you go out and take your ground balls, you take your fly balls off the bat, and you take your batting practice. Pitchers do their sidelines, which is their bullpens, and you go in. You sit for about an hour, hour and a half. Get that uniform on, and play your game. So, basically, like I always say, baseball is a job. You're there from 4 p.m. until about 11 p.m. That's a seven-hour day, not counting extra innings if you need them. And then there is the travel.

The average fan doesn't see what happens prior to the game, the preparation and work the coaches and players put in. Then you have to have the energy to play a three- or four-hour game. And there have been games that have gone for eighteen, even twenty innings.

I would sit on the bench and I would listen to the greats. I had the opportunity to sit on the bench and listen George Brett talk about stuff—"Hey, this guy's doing this" and the like. Hitters, when they're watching the game, they're trying to pick up little things that a hitter does. For instance, some pitchers tip their pitches by digging in their glove. Now, pitchers like to keep their finger out of their glove. Now they've made a new thing on the glove where you can put your finger outside your glove but hidden behind a piece of leather.

I was tipping my pitches that way, with the finger that is outside the glove, one time, and the hitters were getting around on me. Remember Steve Howe, the troubled Yankees pitcher, the pitcher who was busted for drugs several times? Howe was suspended seven times over the course of his 17-year career. He came running to me one day, and he says, "Rusty, you're tipping your pitches over there. The team knows what you're throwing." Then he says, "Don't say I told you," because he was a Yankee and I was with the Royals. But he told me I was tipping my pitches with my finger. They knew what was coming. (Howe is dead now, killed in an automobile accident out in California.)

I think I made about $3 million playing the game. To me, man, it's not about the money. I had a great time and I met a lot of good people. There are some great things that I'm proud of. Just recently I was named one the 100 greatest Royals to ever play. I'm number ninety-eight; that's gratifying to me. I wouldn't mind if I was 100, to be in that class with Frank White, George Brett, Amos Otis, Paul Splittorff, Hal McRae, Bret Saberhagen, and Willie Wilson, among others. It's good company to be in.

season, coaches are usually too busy or preoccupied to look at unsolicited mail. It always pays to be considerate. It also shows the prospective coach you may have a sense of poise. We're again getting back to the necessity for mental toughness. Every coach knows it.

- Try to build relationships with the college coaches you've picked out. Once a high school player decides to try to play baseball at the collegiate level, he should sit down and write letters of interest to head coaches of the colleges where he wants to play. Then send the e-mail with your video link. When you go on the road, usually with your parents, to get a physical look at schools you're interested in, make sure you also pay a visit to the baseball coach. Putting a face to a name helps both you and the coach assess the other. When you do talk with the coach, make sure he gets a sense that you will follow up with him. Even if you decide against a school, make sure to let the coach know your decision if you have developed a rapport with him. Your first choice may not end up being a good fit for you after all, so keeping your bridges intact gives you options. Your current coaches can help, too. If you're unsure where you fit in the college scheme of things, ask your coach what he thinks. He may not be always right—you can still get better after high school—but it's another opinion, and you can't have too many. Also, ask your coach—high school, American Legion, Babe Ruth League, AAU, travel team, wherever you are playing—to contact the coaches of the schools you are interested in. They also may have connections with certain coaches. College coaches trust in other coaches to help guide them in the immense forest of high school ball players.

From 1950 until 2010, the College World Series was played at Johnny Rosenblatt Stadium in Omaha, Nebraska. The 2010 series, the last to be played at Rosenblatt, pitted the University of South Carolina against the University of California, Los Angeles, in the championship round, with South Carolina's Gamecocks taking home the cherished hardware. *Getty Images*

- There are other showcases and tournaments not associated with AAU baseball. Usually the tournament will provide a list of recruiters scheduled to attend—if they don't offer a list, pass. The closer the showcase the better, but it also depends on where you want to go to school. Also, many colleges are now running their own baseball camps. They're not showcases, per se, but colleges do recruit from them.

- There are agencies that will go out and help you find scholarships, for a 10 percent fee. If they get you a $20,000 scholarship annually for each year, that may be worth it. Then again, if you're good enough to land a $20,000 scholarship, you probably would have been able to do so without an agent by following the guidelines above.

Arizona State University's baseball program is considered one of the best in the collegiate ranks, producing nearly 100 future major leaguers over the years. Here Sun Devils pitcher Jason Urquidez is on the mound during the 2005 College World Series against the Florida Gators. *Getty Images*

Chance Ruffin of the Texas Longhorns throws a pitch against the Arizona State Sun Devils during the 2009 College World Series. Another elite college program, the University of Texas has appeared in a record 33 College World Series dating back to 1949, with six titles to its credit.

Managing Scholarships

If one athletic department is offering you a larger scholarship than another school that more closely fits your post-graduate career, consider your choice carefully. Keep in mind that you are much more likely to be doing something other than playing baseeball for a living after school. No matter which school you choose, you'll find that collegiate athletics can be great. The fellowship and friends you'll make will often stay with you for years, way past your playing days, because you will have experienced something in college that a lot of other kids don't get a chance to. That team bond is a strong one, especially if you play together for a few years.

What does it take to keep that scholarship? Good grades. Keep up with your studies. It's why you are in school to begin with.

All scholarships are not created equal. Division III schools are forbidden to offer athletic scholarships, but sometimes the financial aid office can find a little extra for the kid the coach wants. There are many questions that have to be answered regarding scholarships, before you choose a school and after.

Acknowledgments

It would have been impossible to put together this book without the generous help, both in labor and in spirit, of a number of people.

Len Solesky wants to thank his wife, Deborah Ann Solesky, for her encouragement and support; his father, Walter Solesky, for giving the guidance that helps him communicate with young players; and his English teacher at Enfield (Conn.) High School, Raymond Mercik, for his encouragement.

James T. Cain would like to thank Eric Cressey, who contributed the chapter on strength and conditioning, read the manuscript, and offered additional expertise in many areas. Dr. Scott Silverberg, who lent his knowledge on orthopedic medicine, provided a sounding board for questions on physiology. Rob Ackerly gave us an insight into what a physical therapist does and why, and served as model in demonstrating the techniques used to fix, restore, and rehabilitate injured musculature. Our editor at MVP Books, Adam Brunner, did an utterly astounding job, beyond yeoman's work, putting it all together. Thanks go to publisher Josh Leventhal, too, for continued belief in the project.

The young pitchers who agreed to help us include Brian and Michael Pendergast, Colin Sledzik, Connor Ferguson, Austin Young, Alex and David Julian, and Nick Bonofiglio.

We want to thank Kingswood Oxford School in West Hartford, Connecticut, for the generous use of the school's baseball diamond. Athletic Director Garth Adams even turned on the scoreboard for us. Sonya Adams, the school's director of communications, was very helpful in arranging for the video shoot. Along with many other things not associated with the video, Nancie Woodford-Cain catered for us and even came up with a tent so the kids could take shelter on a couple of the hottest days of summer.

Rusty Meacham thanks Bob Shaw, who once beat Sandy Koufax, 1–0, in the 1959 World Series. Shaw taught Meacham all the things that made him successful in the major leagues: a strong work ethic, goals, and a belief in his ability. Other big influences were Mike Easom, his coach at Indian River Community College (now State College); and his high school coaches, Bob Merlano, and Mike Lindgren.

Bruce Curtis thanks Larry Oswald, Andrew Oswald, Phil Royse, Brandon Gluckstal, Ryan Siegelstein, Jerry Sielstein, and Studio Lisa.

Index

About the Authors

Len Solesky was a professional scout for 18 years and a pitching consultant for 9 years, working in the big leagues with the Houston Astros, Tampa Bay Rays, Los Angeles Dodgers, and the Atlanta Braves. Solesky has coached teams in Division I college, high school, American Legion, and AAU, and currently works privately with professional, collegiate, and high school pitchers nationwide. A renowned pitching instructor, Solesky teaches a new approach to pitching that incorporates recent medical advances.

James T. Cain is the co-author of *One-on-One Baseball*, written with former New York Yankees bullpen coach Dom Scala. With a background in journalism, Cain worked as an editor with the Hartford *Courant* for almost 20 years. Continuing a family pitching tradition, Cain's son pitches for Washington University in Saint Louis (His grandfather once pitched against the infamous Black Sox before they were banned from baseball).

Rusty Meacham pitched professionally for almost 20 seasons, with his most successful stretch coming as a reliever with the Kansas City Royals from 1992-1995. Meacham's best season was 1992, when he posted a 2.74 ERA with 64 strikeouts in as many games. Meacham's success has carried over to a career as a pitching coach and teacher, most recently with minor league clubs in the Washington Nationals system.

Bruce Curtis has been a professional photographer for many decades, contributing to such esteemed publications as *Time*, *Life*, and *Sports Illustrated*.